ELEMENTARY DIFFERENTIAL EQUATIONS AND OPERATORS

by

G. E. H. REUTER

LONDON: Routledge & Kegan Paul Ltd

First published 1958
in Great Britain by
Routledge & Kegan Paul Limited
Broadway House, 68–74 Carter Lane
London, EC4V 5EL

Reprinted 1959, 1962, 1964, 1966, 1971

ISBN 0 7100 4342 2

Printed in Great Britain
by Butler & Tanner Limited
Frome and London

ELEMENTARY DIFFERENTIAL EQUATIONS AND OPERATORS

LIBRARY OF MATHEMATICS

edited by

WALTER LEDERMANN

D.Sc., Ph.D., F.R.S.Ed., Professor of Mathematics, University of Sussex

Complex Numbers	W. Ledermann
Differential Calculus	P. J. Hilton
Differential Geometry	K. L. Wardle
Electrical and Mechanical Oscillations	D. S. Jones
Elementary Differential Equations and Operators	G. H. H. Reuter
Fourier and Laplace Transforms	P. D. Robinson
Fourier Series	I. N. Sneddon
Functions of a Complex Variable 2 vols	D. O. Tall
Integral Calculus	W. Ledermann
Introduction to Abstract Algebra	C. R. J. Clapham
Linear Equations	P. M. Cohn
Multiple Integrals	W. Ledermann
Numerical Approximation	B. R. Morton
Partial Derivatives	P. J. Hilton
Principles of Dynamics	M. B. Glauert
Probability Theory	A. M. Arthurs
Sequences and Series	J. A. Green
Sets and Groups	J. A. Green
Solid Geometry	P. M. Cohn
Solutions of Laplace's Equation	D. R. Bland
Vibrating Strings	D. R. Bland
Vibrating Systems	R. F. Chisnell

Preface

THE main purpose of this book is to give a thoroughly elementary account of the 'operational method' for solving linear differential equations with constant coefficients, subject to prescribed initial conditions.

No previous knowledge of differential equations will be demanded from the reader, and accordingly the basic properties of linear differential equations with constant coefficients are treated in the first chapter. This chapter may be found useful even by those readers who do not wish to learn the more specialized technique of operators, treated in the second chapter.

The book is addressed primarily to students of the exact sciences and of engineering, but it deals only with mathematical techniques and does not contain any illustrative examples drawn from other fields (such as mechanics, electric circuit theory, probability theory, and so on). Each reader will no doubt meet problems in his own field of studies to which the mathematical techniques described in this book may usefully be applied.

I am very grateful to Dr. Walter Ledermann for reading the draft with great care and for making many useful suggestions.

Contents

PAGE

CHAPTER II: THE OPERATIONAL METHOD

§ 1 Preliminary Discussion of the Method

§ 2 Practical Instructions for using the Method

CHAPTER ONE

Linear Differential Equations with Constant Coefficients

---

§ 1. The first order equation

1.1 Introduction. The linear differential equation of the first order with constant coefficients is the simplest of the equations which will be treated in this book, and thorough familiarity with its properties will be absolutely essential in everything that follows.

The equation is most conveniently written in the form

$$\frac{dy}{dx}+ay=f(x). \tag{1}$$

Here a is a *constant* (i.e. does not depend on x or y), f is a *given* function of x, and y is an *unknown* function of x which we have to find from (1) by 'solving' this equation. The equation is called 'linear' because it contains only a linear combination of y and dy/dx, as opposed to non-linear combinations such as $y\frac{dy}{dx}$, and first order because it contains only the first derivative dy/dx and not the higher derivatives d^2y/dx^2, . . .

One instance of (1) will certainly be known to the reader: the equation

$$\frac{dy}{dx}=f(x). \tag{2}$$

Here we have to find a function y whose derivative equals the given function f, and this is simply the problem of integrating f. Thus y must be an *indefinite integral* of f:

$$y=\int f(x)dx+C, \tag{3}$$

where C can be any constant. The general solution of (2)

therefore involves one *arbitrary constant* whose value is at our disposal. In practice it is often not the general solution which is needed, but rather some special solution, for instance that solution for which y has some given value y_0 when x has some given value x_0. For example, if $dy/dx=3x^2-2$ then $y=x^3-2x+C$, but if we also require that $y=1$ when $x=2$ then we must take $C=-3$. Generally, if an *initial condition*:

$$y=y_0 \text{ when } x=x_0,$$

is imposed then the value of the arbitrary constant C in (3) can be determined. The required solution is then most simply written in terms of a *definite integral* of f, as

$$y=y_0+\int_{x_0}^{x} f(\xi)d\xi. \tag{4}$$

Another familiar special case of (1) occurs when the right-hand member $f(x)$ is zero, so that the equation reduces to

$$\frac{dy}{dx}+ay=0. \tag{5}$$

Dividing through by y and noting that

$$\frac{1}{y}\frac{dy}{dx}=\frac{d(\log y)}{dx}$$

we can then write

$$d(\log y)/dx=-a,$$

whence

$$\log y=-ax+C,\ y=C'e^{-ax},$$

where C is an arbitrary constant and $C'=e^C$. The preceding argument is, however, inadequate because it ignores the possibilities that y may be zero (then we must not divide by y) or, even worse, that y may be negative (then $\log y$ is not defined). It would not be hard to patch up the argument and to arrive at the complete solution to (5), but for us it will be better to make a fresh start: we shall give another method of treating (5) which not only leads to the complete solution of (5) but can also be used to solve the more general equation (1).

1.2 The integrating factor. We now want to show that the general solution of (5) is

$$y=Ke^{-ax}, \tag{6}$$

where K is an arbitrary constant. Now (6) is equivalent to

$$ye^{ax}=\text{constant},$$

which in turn is equivalent to

$$\frac{d}{dx}(ye^{ax})=0,$$

$$e^{ax}\left(\frac{dy}{dx}+ay\right)=0;$$

since e^{ax} is never zero, the last equation is equivalent to (5). Thus (5) and (6) are equivalent: every choice of K in (6) gives a solution of (5), and every solution of (5) has the form (6) for some value of K.

The success of this method is due to the fact that

$$e^{ax}\left(\frac{dy}{dx}+ay\right)$$

is an 'exact derivative', i.e. is the derivative of ye^{ax}. The same fact can be used to solve the general equation (1),

$$\frac{dy}{dx}+ay=f(x);$$

if we multiply by e^{ax} we obtain

$$\frac{d}{dx}(ye^{ax})=e^{ax}f(x),$$

and then we can integrate both sides to obtain

$$ye^{ax}=\int e^{ax}f(x)dx+C,$$

$$y=e^{-ax}\int e^{ax}f(x)dx+Ce^{-ax}. \tag{7}$$

Formula (7) gives the general solution; if an initial condition has to be satisfied, say $y=y_0$ when $x=x_0$, the formula would read

$$y=e^{-ax}\int_{x_0}^{x} e^{a\xi}f(\xi)d\xi+y_0e^{-a(x-x_0)}. \tag{8}$$

In practice it is hardly worth while to try to memorize these two formulae: one need only remember the method by which they were obtained.

Example 1: $$\frac{dy}{dx}-3y=\sin x.$$

Multiply by e^{-3x} in order to bring the equation into the form

$$\frac{d}{dx}(ye^{-3x})=e^{-3x}\sin x,$$

and then integrate to obtain

$$ye^{-3x}=\int e^{-3x}\sin x\,dx+C$$

$$=-\frac{1}{10}e^{-3x}(\cos x+3\sin x)+C.$$

Hence the general solution is

$$y=-\frac{1}{10}(\cos x+3\sin x)+Ce^{3x},$$

where C is an arbitrary constant.

Example 2: *Solve* $\frac{dy}{dx}+y=x^{-1}$ *for* $x>0$, *with the initial condition* $y=0$ *when* $x=1$.

After multiplying by e^{x} we get

$$\frac{d}{dx}(ye^{x})=x^{-1}e^{x}.$$

Integrating this and making use of the initial condition, we get

$$ye^{x}=\int_1^x \xi^{-1}e^{\xi}d\xi$$

and hence the required solution is

$$y=e^{-x}\int_1^x \xi^{-1}e^{\xi}d\xi.$$

It will be noticed that our formula for the solution contains an integral which cannot be expressed in terms of 'elementary' functions (polynomials, exponentials, logarithms, sines and cosines). This will often happen and when it does we must be content to accept such formulae for a solution. In practical problems one is in any case likely to need numerical values for the solution, and from this standpoint the 'known' functions are those whose values have been tabulated: amongst these occur not only the 'elementary' functions mentioned above but many others (including, as it happens, the integral $\int_1^x \xi^{-1}e^{\xi}d\xi$ which occurs in Example 2). We cannot of course expect that the solution will always involve integrals which have already been tabulated, and it may sometimes be necessary to use numerical methods of integration (such as Simpson's Rule).

The work of this section may conveniently be summarized in the following *rule for solving the first order equation* (1):

Multiply by e^{ax} to bring the equation into the form

$$\frac{d}{dx}(ye^{ax})=e^{ax}f(x).$$

Then integrate both sides; initial conditions can be fitted at this stage by using appropriate definite integrals (as in Example 2).

It is usual to call e^{ax} an *integrating factor* for equation (1) because multiplication by this factor enables us to solve the equation by simple integration.

1.3 The form of the general solution. We return briefly to the general formula (7),

$$y=e^{-ax}\int e^{ax}f(x)dx+Ce^{-ax},$$

for the solution of $dy/dx+ay=f(x)$. Taking $C=0$ we see that

$$y=e^{-ax}\int e^{ax}f(x)dx$$

is a *particular solution*, and that the general solution is obtained by adding Ce^{-ax}. Now Ce^{-ax} is the general solution of the *reduced equation* $dy/dx+ay=0$; it is usually called the *complementary function*. Thus we can say:

'General solution = particular solution
plus complementary function.'

We shall find that a similar assertion may be made for linear equations of higher order and will then be very useful in finding the general solution.

§ 2. The second order equation

2.1. The reduced equation. The general second order equation reads

$$\frac{d^2y}{dx^2}+a\frac{dy}{dx}+by=f(x), \tag{1}$$

but we shall first treat the *reduced equation* in which the right-hand member is zero, i.e.

$$\frac{d^2y}{dx^2}+a\frac{dy}{dx}+by=0. \tag{2}$$

Here a and b are supposed to be *constants*; they must not depend on x.

The first order equation

$$\frac{dy}{dx}+ay=0,$$

analogous to (2), has solutions Ce^{-ax}. We may therefore expect (2) to have exponential solutions also, and indeed we can easily find such solutions by substituting $y=e^{kx}$ into (2). This gives

$$k^2e^{kx}+ake^{kx}+be^{kx}=0,$$

and so $y=e^{kx}$ will be a solution provided that

$$k^2+ak+b=0.$$

This quadratic has two roots, k_1 and k_2, and so we obtain two solutions, e^{k_1x} and e^{k_2x}, of (2). Now whenever we know two solutions, say y_1 and y_2, of (2), then $C_1y_1+C_2y_2$ will also be a solution for arbitrary values of the constants C_1 and C_2; for if $y=C_1y_1+C_2y_2$ then

$$\frac{d^2y}{dx^2}+a\frac{dy}{dx}+by$$

$$=\left(C_1\frac{d^2y_1}{dx^2}+C_2\frac{d^2y_2}{dx^2}\right)+a\left(C_1\frac{dy_1}{dx}+C_2\frac{dy_2}{dx}\right)+b(C_1y_1+C_2y_2)$$

$$=C_1\left(\frac{d^2y_1}{dx^2}+a\frac{dy_1}{dx}+by_1\right)+C_2\left(\frac{d^2y_2}{dx^2}+a\frac{dy_2}{dx}+by_2\right)$$

$$=C_1(0)+C_2(0)=0.$$

We have therefore found a solution

$$y=C_1e^{k_1x}+C_2e^{k_2x}$$

which contains two arbitrary constants. This is quite satisfactory except when k_1 and k_2 happen to be equal; then we merely get $y=(C_1+C_2)e^{k_1x}$, i.e. an arbitrary multiple of e^{k_1x} only, and we may expect that there is a second solution which we must still find. Also we cannot be sure (even when k_1 and k_2 are not equal) that we have found all possible solutions. We therefore make a fresh start, using a better method which will not only resolve our present difficulties but can also be used to deal with the general equation (1).

The new method depends on 'factorizing' the left-hand side of (2) in a way which is perhaps made clearer by adopting a new notation. Let us write 'D' for 'd/dx', so that $\frac{dy}{dx}=\frac{d}{dx}y$ will be denoted by Dy. Then

$$\frac{d^2y}{dx^2}=\frac{d}{dx}\left(\frac{dy}{dx}\right)=D(Dy),$$

and naturally we shall abbreviate $D(Dy)$ to D^2y. Equation (2) can now be written as

$$(D^2+aD+b)y=0,$$

which at once suggests that we should factorize the expression in brackets as $(D-k_1)(D-k_2)$, just as we would factorize k^2+ak+b as $(k-k_1)(k-k_2)$. Since 'D' is not a number but a symbol denoting 'differentiation', the reader will be quite justified in doubting whether such a factorization is legitimate. It is in fact legitimate because D satisfies the usual arithmetical laws relating to addition, subtraction and multiplication. To put it less abstractly: once we have agreed that D and D^2 shall stand for d/dx and d^2/dx^2, a reasonable interpretation for $(D-k_1)(D-k_2)y$ is

$$(D-k_1)\left(\frac{dy}{dx}-k_2y\right)=\frac{d}{dx}\left(\frac{dy}{dx}-k_2y\right)-k_1\left(\frac{dy}{dx}-k_2y\right)$$

$$=\frac{d^2y}{dx^2}-k_2\frac{dy}{dx}-k_1\frac{dy}{dx}+k_1k_2y$$

$$=\frac{d^2y}{dx^2}-(k_1+k_2)\frac{dy}{dx}+k_1k_2y$$

$$=\frac{d^2y}{dx^2}+a\frac{dy}{dx}+by,$$

because $k_1+k_2=-a$ and $k_1k_2=b$, k_1 and k_2 being the roots of $k^2+ak+b=0$.

We have now written (2) as $(D-k_1)(D-k_2)y=0$. If we put $(D-k_2)y=z$, for the moment, this becomes $(D-k_1)z=0$, or $dz/dx-k_1z=0$. This gives $z=Ce^{k_1x}$, and now we have

$$\frac{dy}{dx}-k_2y=z=Ce^{k_1x},$$

which is a first order linear equation for y. Solving it by the standard method we get

$$\frac{d}{dx}(ye^{-k_2x})=Ce^{(k_1-k_2)x},$$

$$ye^{-k_2x}=C\int e^{(k_1-k_2)x}dx+C'.$$

If $k_1 \neq k_2$, the integral on the right is

$$\frac{C}{k_1-k_2}e^{(k_1-k_2)x};$$

if $k_1=k_2$, it is simply $\int Cdx=Cx$. We have now proved that the most general solution of (2) is given by

$$y=\frac{C}{k_1-k_2}e^{k_1x}+C'e^{k_2x} \quad \text{if } k_1 \neq k_2,$$

$$y=Cxe^{k_1x}+C'e^{k_1x} \quad \text{if } k_1=k_2.$$

Note that $C/(k_1-k_2)$ is an *arbitrary* constant: we can give it whatever value we wish by suitably choosing C. Thus the general solution of (2) is an *arbitrary linear combination* of

$$e^{k_1x} \text{ and } e^{k_2x} \quad \text{if } k_1 \neq k_2,$$

$$e^{k_1x} \text{ and } xe^{k_1x} \text{ if } k_1=k_2.$$

Incidentally we have found the missing second solution when $k_1=k_2$: it is xe^{k_1x}.

One further point needs discussion: what happens if the roots of $k^2+ak+b=0$ are *complex*? In practice a and b will certainly be real, so that k_1 and k_2 will be complex conjugates, say

$$k_1=\alpha+i\omega,\ k_2=\alpha-i\omega.$$

The corresponding solutions will be

$$e^{(\alpha\pm i\omega)x}=e^{\alpha x}(\cos\omega x\pm i\sin\omega x).$$

Any linear combination of $e^{(\alpha+i\omega)x}$ and $e^{(\alpha-i\omega)x}$ can therefore be rewritten as a combination of $e^{\alpha x}\cos\omega x$ and $e^{\alpha x}\sin\omega x$, say

$$e^{\alpha x}(A\cos\omega x+B\sin\omega x);$$

it is usually preferable to write the solution in this form because the constants A and B will then be real for a real-valued solution.

Example 1: $\qquad (D^2+2D+2)y=0.$

Here $k^2+2k+2=0$, so $(k+1)^2+1=0$ and $k+1=\pm i$, or $k=-1\pm i$. Thus $\alpha=-1$, $\omega=1$, and the solution is

$$y=e^{-x}(A\cos x+B\sin x).$$

Perhaps the reader who feels any doubt about the use made of complex numbers in this calculation may be slightly reassured when we have checked that this is a solution. We have

$$\begin{aligned} Dy &= -e^{-x}(A\cos x+B\sin x)+e^{-x}(-A\sin x+B\cos x)\\ &=e^{-x}[(B-A)\cos x-(A+B)\sin x], \end{aligned}$$

and similarly

$$D^2y=e^{-x}[-2B\cos x+2A\sin x].$$

Hence $(D^2+2D+2)y$ is a linear combination of $e^{-x}\cos x$ and $e^{-x}\sin x$, with coefficients

$$\begin{aligned} -2B+2(B-A)+2A&=0,\\ 2A-2(A+B)+2B&=0; \end{aligned}$$

that is to say, $(D^2+2D+2)y=0$.

2.2 The general equation. The 'factorization' method, used above to solve the reduced equation $(D^2+aD+b)y=0$, can be used equally well to solve the general equation (1):

$$(D^2+aD+b)y=f(x),$$

when the right-hand member $f(x)$ is not zero. Putting $(D-k_2)y = z$ as before, we now have

$$(D-k_1)z=f(x).$$

After solving this first order linear equation for z, we can then solve

$$(D-k_2)y= z$$

for y.

Unfortunately this method usually involves rather clumsy calculations because of the integrations needed for solving two first order linear equations, and the method is *not* recommended for practical use. If we did use it, we should find that the solution consists of two parts: one part, containing two arbitrary constants of integration, coincides with the solution of the reduced equation (see p. 8) and is called the *complementary function*; the other part depends on $f(x)$ and is a *particular solution*. Thus *the general solution of*

$$(D^2+aD+b)y=f(x)$$

can be found by taking one particular solution and adding to it the complementary function (i.e. the general solution of the reduced equation $(D^2+aD+b)y=0$, which contains two arbitrary constants).

In view of the practical importance of the above statement, we will give a direct verification. Let y_1 be a particular solution of (1) and let y_2 be any solution of the reduced equation. Then if $y=y_1+y_2$ we have

$$\begin{aligned}(D^2+aD+b)y&=(D^2+aD+b)y_1+(D^2+aD+b)y_2\\&=f(x)+0=f(x),\end{aligned}$$

so that y is also a solution of (1). Conversely, if y is any solution of (1) and if we put $y-y_1=y_2$, then

$$\begin{aligned}(D^2+aD+b)y_2&=(D^2+aD+b)y-(D^2+aD+b)y_1\\&=f(x)-f(x)=0,\end{aligned}$$

so that y_2 satisfies the reduced equation; this shows that y is necessarily of the form y_1+y_2.

Now it is often possible to find a particular solution fairly easily without having to use the general method explained at the beginning of this section. In particular, this can always be done when $f(x)$ is a *polynomial*, an *exponential*, or a combination of *sines and cosines*. The next three sections will show how it can be done in these three special cases.

2.3 Particular solution: polynomial $f(x)$. If $f(x)$ is a polynomial, we can always find a particular solution which is also a polynomial by substituting into the equation a trial polynomial (with undetermined coefficients) and then equating coefficients.

Example 2: $$(D^2-7D-5)y=x^3-1.$$

We try a particular solution of the form $y=px^3+qx^2+rx+s$. Then

$$Dy=3px^2+2qx+r,\ D^2y=6px+2q,$$

and we must have

$$6px+2q-7(3px^2+2qx+r)-5(px^3+qx^2+rx+s)=x^3-1.$$

Equating the coefficients of x^3, x^2, x and 1, we find that

$$-5p=1,\ p=-\frac{1}{5};$$

$$-21p-5q=0,\ q=-\frac{21}{5}p=\frac{21}{25};$$

$$6p-14q-5r=0,\ -\frac{6}{5}-\frac{294}{25}=5r,\ r=-\frac{324}{125};$$

$$2q-7r-5s=-1,\ \frac{42}{25}+\frac{2268}{125}+1=5s,\ s=\frac{2603}{625}.$$

Hence a particular solution is

$$y=\tfrac{1}{625}(-125x^3+525x^2-1620x+2603).$$

Notice that the trial solution was taken to be a polynomial of the same degree as the right-hand side x^3-1 of the equation. This will suffice unless the coefficient b (of y on the left-hand side) is zero; in this case the degree of the trial polynomial must be increased by 1. The reason for this is that when $b=0$ the constant term s in the trial solution $y=\ldots+s$ will disappear when we form $(D^2+aD+0)y$, and will not be available when we equate coefficients. We therefore need another coefficient at the start of the trial solution. If a, the coefficient of Dy, is also zero then the equation is simply $D^2y=f(x)$ and can be solved by integrating twice.

2.4 Particular solution: exponential $f(x)$. Another case in which the form of a particular solution can be predicted is when $f(x)$ is an exponential, so that the equation reads

$$(D^2+aD+b)y=Me^{mx}.$$

In general one can then find a particular solution of the form $y=Ae^{mx}$, and the constant A can be found by substituting into the equation. We must have

$$Am^2e^{mx}+a.Ame^{mx}+b.Ae^{mx}=Me^{mx},$$
$$A(m^2+am+b)=M.$$

This fixes A, *unless* $m^2+am+b=0$. In this case, $k=m$ is a root of the equation $k^2+ak+b=0$: thus e^{mx} occurs in the complementary function, and naturally the result of substituting $y=Ae^{mx}$ into the left-hand side is zero. We must then try $y=Axe^{mx}$ and find A as before by substituting. Finally it may happen that both e^{mx} and xe^{mx} already occur in the complementary function, i.e. that $k=m$ is a *double* root of $k^2+ak+b=0$. We must then try $y=Ax^2e^{mx}$ and find A by substituting. (We have not attempted to explain why multiples of xe^{mx} or x^2e^{mx} should occur in the exceptional cases mentioned above. The reader may either accept this as an experimental fact or may derive it for himself by applying the general method of solution explained in § 2.2.)

Example 3: $\qquad (D^2-3D+2)y=5e^{3x}.$

Substituting $y=Ae^{3x}$, we get

$$9Ae^{3x}-3.3Ae^{3x}+2Ae^{3x}=5e^{3x},$$
$$2Ae^{3x}=5e^{3x},\ A=\frac{5}{2}.$$

Hence $y=\frac{5}{2}e^{3x}$ is a particular solution.

Example 4: $\qquad (D^2-1)y=3e^{-x}.$

Here e^{-x} already occurs in the complementary function; it is useless to substitute Ae^{-x}, and we must try Axe^{-x} instead. If $y=Axe^{-x}$, then

$$Dy=A(-x+1)e^{-x},\ D^2y=A(x-2)e^{-x}.$$

So we must have

$$A(x-2)e^{-x}-Axe^{-x}=3e^{-x}.$$

The terms in xe^{-x} cancel as they should (this is always a useful check), and we are left with

$$-2Ae^{-x}=3e^{-x},\ A=-\frac{3}{2},$$

so that $y=-\frac{3}{2}xe^{-x}$ is a particular solution.

Example 5: $(D^2-2D+1)y=e^x.$

Here $k^2-2k+1=0$ has $k=1$ as a double root, and therefore both e^x and xe^x occur in the complementary function. Thus we have the worst possible case and must try $y=Ax^2e^x$. This gives

$$Dy=A(x^2+2x)e^x,\ D^2y=A(x^2+4x+2)e^x.$$

Substituting and cancelling the common factor e^x, we get

$$A(x^2+4x+2)-2A(x^2+2x)+Ax^2=1.$$

The terms in x^2 and x cancel (again a useful check), and this simply leaves $2A=1$, $A=\frac{1}{2}$; $y=\frac{1}{2}x^2e^x$ is a particular solution.

2.5 Particular solution: trigonometric $f(x)$. Finally, we deal with the case when $f(x)$ is a multiple of $\cos nx$ or $\sin nx$, or more generally of the form

$$f(x)=P\cos nx+Q\sin nx.$$

There are several methods, of which we give two. First, we may try a solution of the same form as $f(x)$, namely

$$y=A\cos nx+B\sin nx,$$

and find A and B by substituting in the equation.

Example 6: $(D^2-2D-5)y=2\cos 3x-\sin 3x.$

Substituting $y=A\cos 3x+B\sin 3x$, we have

$$(-9A\cos 3x-9B\sin 3x)-2(-3A\sin 3x+3B\cos 3x)$$
$$-5(A\cos 3x+B\sin 3x)=2\cos 3x-\sin 3x.$$

Equate the coefficients of cos $3x$ and sin $3x$ to zero; this gives

$$-14A-6B=2,\ 6A-14B=-1,$$

and hence

$$(14^2+6^2)A=2(-14)-1(6),\quad 232A=-34,\ A=-17/116,$$
$$(14^2+6^2)B=2(-6)-1(-14),\ 232B=2,\quad B=1/116.$$

Thus we have as a particular solution

$$y=\frac{1}{116}(-17\cos 3x+\sin 3x).$$

The preceding method will fail when cos nx and sin nx already occur in the complementary function, i.e. when the equation is

$$(D^2+n^2)y=P\cos nx+Q\sin nx.$$

The trial solution must then taken to be

$$y=x(A\cos nx+B\sin nx);$$

see Problem 12.

A second method for finding a particular solution of

$$(D^2+aD+b)y=P\cos nx+Q\sin nx$$

is based on the fact that cos nx and sin nx can be expressed in terms of *complex exponentials*. For instance cos nx and sin nx are the real and imaginary parts of e^{inx}, and more generally $P\cos nx+Q\sin nx$ is the real part of

$$(P-iQ)(\cos nx+i\sin nx)=(P-iQ)e^{inx}.$$

(We shall assume that P and Q, as well as the coefficients a and b, are real.) Now suppose that we have found a complex-valued solution y of

$$(D^2+aD+b)y=f(x),$$

where $f(x)=f_1(x)+if_2(x)$ is also complex-valued. Writing $y=y_1+iy_2$, where y_1 and y_2 are the real and imaginary parts of y, we have

$$(D^2+aD+b)y_1+i(D^2+aD+b)y_2=f_1(x)+if_2(x).$$

Equating real parts,

$$(D^2+aD+b)y_1=f_1(x),$$

and equating imaginary parts,

$$(D^2+aD+b)y_2=f_2(x).$$

Thus if y is a particular solution when the right-hand side is (x), then $\mathbf{R}y$ and $\mathbf{I}y$ are solutions when the right-hand side is $\mathbf{R}f(x)$ and $\mathbf{I}f(x)$, respectively. To find a solution when the right-hand side is K cos nx or K sin nx, we may therefore first find a solution when the right-hand side is Ke^{inx}, and then take its real or imaginary part; more generally when we have $P\cos nx + Q\sin nx$ on the right, replace it by $(P-iQ)e^{inx}$, find a solution, and then take its real part.

We illustrate the method for the same equation as in Example 6.

Example 7: $\qquad (D^2-2D-5)y=2\cos 3x-\sin 3x.$

The right-hand side is the real part of $(2+i)e^{3ix}$. To find a particular solution when the right-hand side is $(2+i)e^{3ix}$, substitute $y=Ae^{3ix}$. Then after cancelling e^{3ix} we get

$$A[(3i)^2-2(3i)-5]=2+i,$$

$$A=-\frac{2+i}{14+6i}=-\frac{(2+i)(14-6i)}{14^2+6^2}=-\frac{34+2i}{232}=-\frac{1}{116}(17+i).$$

Taking the real part of the corresponding solution Ae^{3ix}, we obtain the required solution

$$y=\mathbf{R}\left\{-\frac{1}{116}(17+i)(\cos 3x+i\sin 3x)\right\}$$

$$=-\frac{1}{116}(17\cos 3x-\sin 3x).$$

A similar method applies in the exceptional case when cos nx and sin nx already occur in the complementary functions

Example 8: $\qquad (D^2+4)y=3\sin 2x.$

Here we shall first find a solution when the right-hand side is $3e^{2ix}$ and then take its imaginary part. Since e^{2ix} already occur. in the complementary function we must take a trial solution of the form $y=Axe^{2ix}$. This gives

$$Dy=A(2ix+1)e^{2ix},\ D^2y=A(-4x+4i)e^{2ix}.$$

Substituting and cancelling e^{2ix},

$$A(-4x+4i)+4Ax=3;\ A=\frac{3}{4i}=-\frac{3}{4}i.$$

Hence the required solution is

$$y=\mathbf{I}\{-\tfrac{3}{4}ixe^{2ix}\}=-\tfrac{3}{4}x\cos 2x.$$

2.6 Particular solution: some further cases. The methods of the preceding three sections can be extended to deal with more complicated forms of right-hand side $f(x)$. We shall merely state what procedures can be used: examples can be found amongst the problems at the end of this chapter.

(i) If $f(x)$ has the form

$$e^{mx}(P \cos nx + Q \sin nx),$$

either substitute a trial solution of similar form, viz.

$$y = e^{mx}(A \cos nx + B \sin nx),$$

or replace the right-hand side by

$$(P - iQ)e^{(m+in)x},$$

find a solution, and take its real part. As usual a factor x must be inserted if $e^{mx} \cos nx$ and $e^{mx} \sin nx$ occur in the complementary function. Unless the reader has strong objections to the use of complex numbers, he is advised to use the second of the above methods.

(ii) If $f(x)$ consists of a polynomial multiplied by an exponential factor, try a solution of the same form; the polynomial occurring in the trial solution should have the same degree as that occurring in $f(x)$, but its degree must be increased in exceptional cases. For instance suppose that $f(x) = (x^2 + 1)e^{2x}$. Then take $y = (px^2 + qx + r)e^{2x}$ as trial solution; but if e^{2x} occurs in the complementary function take $y = (px^3 + qx^2 + rx)e^{2x}$, and if xe^{2x} also occurs take $y = (px^4 + qx^3 + rx^2)e^{2x}$.

(iii) Similarly, if $f(x)$ is a polynomial multiplied by $(P \cos nx + Q \sin nx)$, either try a solution of the same form or (better) replace the right-hand side by the same polynomial multiplied by $(P - iQ)e^{inx}$, find a solution as in (ii) above, and then take its real part.

(iv) It may happen that $f(x)$ is the sum of several terms, each of which could be treated by substituting an appropriate trial solution. Clearly we can deal with such an $f(x)$ by substituting an appropriate combination of terms. For instance if $f(x) = e^x - 2e^{-3x} + x^2$, substitute $y = Ae^x + Be^{-3x} + px^2 + qx + r$ and hence determine A, B, p, q and r.

2.7 Arbitrary constants and initial conditions. The *general solution* of an equation

$$(D^2+aD+b)y=f(x)$$

has the form $y=y_1+y_2$, where y_1 is a particular solution and y_2 is the complementary function containing *two arbitrary constants*. One often wants not the general solution, but rather the special solution which satisfies given *initial conditions*: y and Dy are to have given values at some given value of x. The appropriate values of the arbitrary constants can then always be found by substitution, as in the example below.

Example 9:

Find the solution of $(D^2-4D-5)y=x^2$ *for which* $y=1$ *and* $Dy=-1$ *when* $x=0$.

The quadratic $k^2-4k-5=0$ has roots $k=5, -1$, so that the complementary function is $Ae^{5x}+Be^{-x}$. As a particular solution, try $y=px^2+qx+r$. This gives

$$2p-4(2px+q)-5(px^2+qx+r)=x^2;$$

equating coefficients,

$$-5p=1, \quad -8p-5q=0, \quad 2p-4q-5r=0,$$

and hence

$$p=-\frac{1}{5}; \; q=-\frac{8}{5}p=\frac{8}{25}; \; 5r=2p-4q=-\frac{2}{5}-\frac{32}{25}, \; r=-\frac{42}{125}.$$

So the general solution is

$$y=-\frac{1}{5}x^2+\frac{8}{25}x-\frac{42}{125}+Ae^{5x}+Be^{-x}.$$

For this solution we have

$$\left.\begin{aligned} y&=-\frac{42}{125}+A+B, \\ Dy&=\frac{8}{25}+5A-B, \end{aligned}\right\} \text{when } x=0.$$

To fit the initial conditions ($y=1$, $Dy=-1$, when $x=0$), we must have

$$-\frac{42}{125}+A+B=1,$$
$$\frac{8}{25}+5A-B=-1.$$

Adding, we get $-\frac{2}{125}+6A=0$, $A=\frac{1}{375}$; then

$$B=1+5A+\frac{8}{25}=1+\frac{1}{75}+\frac{8}{27}=\frac{4}{3}.$$

Thus the required solution is

$$y=-\frac{1}{5}x^2+\frac{8}{25}x-\frac{42}{125}+\frac{1}{375}e^{5x}+\frac{4}{3}e^{-x}.$$

The reader will notice that even a simple example may involve quite heavy calculations because one has first to find the general solution and then to fit the arbitrary constants to the given initial conditions. The arithmetical work can be lightened by using the *operational method*, to be presented in Ch. II; this method is specially designed for finding the solution satisfying given initial conditions, and at the same time will also give the general solution if this is wanted.

2.8 Recapitulation. We shall now give a brief summary of the methods developed in the preceding sections for finding the general solution of

$$(D^2+aD+b)y=f(x).$$

The reader will recall that the general solution is the sum of any one particular solution and of the complementary function; the latter is the solution of the 'reduced equation'

$$(D^2+aD+b)y=0$$

and contains two arbitrary constants. The following *routine for finding the general solution* is recommended:

(i) *Find the roots, k_1 and k_2, of $k^2+ak+b=0$.*

(ii) *Write down the complementary function; this is*
$C_1e^{k_1x}+C_2e^{k_2x}$ if k_1 and k_2 are real and unequal,
$C_1e^{k_1x}+C_2xe^{k_1x}$ if k_1 and k_2 are real and equal,
$e^{\alpha x}(C_1 \cos \omega x+C_2 \sin \omega x)$ if k_1 and k_2 are complex conjugates $\alpha \pm i\omega$;
C_1 and C_2 are arbitrary constants.

(iii) *Find a particular solution. This can be done by substitut-*

ing a trial solution with undetermined coefficients in the following cases.

(iii*a*) *$f(x)$ is a polynomial of degree n. Try a polynomial of the same degree, $y=px^n+qx^{n-1}+\ldots$; if $b=0$, try a polynomial of degree $(n+1)$, $y=px^{n+1}+qx^n+\ldots$.*

(iii*b*) *$f(x)$ an exponential, say $f(x)=Me^{mx}$. Try $y=Ae^{mx}$; this succeeds if m does not equal k_1 or k_2 and gives the particular solution*

$$y=\frac{M}{m^2+am+b}e^{mx}.$$

If $k_1 \neq k_2$ and m equals k_1 or k_2, i.e. if e^{mx} occurs in the complementary function but xe^{mx} does not, try $y=Axe^{mx}$. If $k_1=k_2=m$, i.e. if both e^{mx} and xe^{mx} occur in the complementary function, try $y=Ax^2e^{mx}$.

(iii*c*) *$f(x)$ is a combination of $\cos nx$ and $\sin nx$. If $f(x)=K\cos nx$ (or $K\sin nx$), replace $f(x)$ by Ke^{inx}, find a solution as in* (iii*b*) *and then take its real part (or imaginary part). If $f(x)=P\cos nx+Q\sin nx$, replace $f(x)$ by $(P-iQ)e^{inx}$, find a solution as in* (iii*b*) *and then take its real part. (For another method, not using complex numbers, see* § 2.5.)

(iii*d*) *$f(x)$ is a polynomial multiplied by e^{mx} or by $P\cos nx+Q\sin nx$: see* § 2.6.

(iii*e*) *If $f(x)$ is not of one of the above types (nor a sum of several terms, each of one of these types) use the 'factorization' method,* § 2.2: *see Example* 10 *below.*

(iv) *Add the complementary function, found in step* (ii), *to the particular solution. If initial conditions have to be satisfied, the arbitrary constants can now be determined, as in Example* 9 (§ 2.7).

The following example illustrates case (iii*e*) above, when a particular solution can not easily be found by guessing its form and substituting a trial solution.

Example 10: *Find a particular solution of*

$$(D^2-2D+1)y=1/(1+e^x).$$

Since the left-hand side can be written as $(D-1)^2y$, we shall put $(D-1)y=z$, so that

$$(D-1)z=1/(1+e^x).$$

Multiply this first order equation by the integrating factor e^{-x}, which gives $D(ze^{-x})=e^{-x}/(1+e^x)=e^{-2x}/(1+e^{-x})$. Hence

$$ze^{-x}=\int\frac{e^{-2x}}{1+e^{-x}}dx.$$

Putting $e^{-x}=u$, the integral becomes

$$-\int\frac{u}{1+u}du=-\int\left(1-\frac{1}{1+u}\right)du=-u+\log(1+u);$$

we may ignore the constant of integration because we are only looking for a particular solution. So we may take

$$ze^{-x}=-e^{-x}+\log(1+e^{-x}).$$

Now we must solve $(D-1)y=z$, or $D(ye^{-x})=ze^{-x}$.
This gives

$$ye^{-x}=e^{-x}+\int\log(1+e^{-x})dx,$$

$$y=1+e^x\int\log(1+e^{-x})dx.$$

Note that this method will always require *two* integrations. In the above example, the first integration could be carried out explicitly (but the second could not); in general, however, neither integration can be done explicitly so that one obtains a rather awkward formula containing a repeated integral. Once again the use of the operational method (Ch. II) will improve matters because it gives solutions involving at worst single (and not repeated) integrals.

§ 3. Equations of higher order and systems of first order equations

3.1 The nth order equation. The general linear equation with constant coefficients,

$$\frac{d^ny}{dx^n}+a_1\frac{d^{n-1}y}{dx^{n-1}}+\ldots+a_ny=f(x), \tag{1}$$

can be treated by straightforward extensions of the methods used for second order equations. We again associate with (1) the reduced equation

$$\frac{d^n y}{dx^n}+\ldots+a_n y=0, \tag{2}$$

and begin by treating this. It will be helpful to extend the 'D' notation by writing D^3y for d^3y/dx^3, and so on.

Our knowledge of second order equations suggests that we start by looking for exponential solutions of (2). We find at once, by substituting in (2), that $y=e^{kx}$ is a solution provided that

$$k^n+a_1k^{n-1}+\ldots+a_n=0, \tag{3}$$

In general this equation has n distinct roots $k_1, k_2, \ldots, k_n$, and then we get n distinct solutions $e^{k_1x}, \ldots, e^{k_nx}$. When (3) has multiple roots, however, we get less than the expected quota of n distinct solutions and we must look for further solutions. Suppose then that $k=k_1$ is an r-fold root of (3); this means that the left-hand side of (3) can be written as $P(k)(k-k_1)^r$, where $P(k)$ is a polynomial which does not contain the factor $k-k_1$. We now assert that the r functions

$$e^{k_1x},\ xe^{k_1x},\ x^2e^{k_1x},\ \ldots,\ x^{r-1}e^{k_1x} \tag{4}$$

are all solutions of (2). To prove this we write (2) as

$$(D^n+a_1D^{n-1}+\ldots+a_n)y=0$$

and then, by factorizing the left-hand side, write this as

$$P(D)(D-k_1)^r y=0;$$

it will now suffice to prove that the r functions in (4) all satisfy the simpler equation $(D-k_1)^r y=0$. Now these functions all have the form $e^{k_1x}u$, where u is a power of x, and we have

$$(D-k_1)(e^{k_1x}u)=D(e^{k_1x}u)-k_1e^{k_1x}u=e^{k_1x}Du,$$
$$(D-k_1)^2(e^{k_1x}u)=(D-k_1)(e^{k_1x}Du)=e^{k_1x}D(Du)=e^{k_1x}D^2u;$$

by repeating this argument we finally obtain

$$(D-k_1)^r(e^{k_1x}u)=e^{k_1x}D^ru=0$$

because $D^ru=0$ when $u=1, x, x^2, \ldots, x^{r-1}$. This completes the proof that the functions (4) are solutions of (2). When we have treated each root of (3) in this way, allowing for its multiplicity, we can obtain the full quota of n distinct solutions. Let the roots be $k_1, k_2, \ldots, k_j$ with multiplicities $r_1, r_2, \ldots, r_j$, so that $r_1+r_2+\ldots+r_j=n$. Then the n functions

$$\begin{aligned} &e^{k_1x}, xe^{k_1x}, \ldots, x^{r_1-1}e^{k_1x}; e^{k_2x}, \ldots, x^{r_2-1}e^{k_2x}; \\ &\qquad \ldots \qquad ; e^{k_jx}, xe^{k_jx}, \ldots, x^{r_j-1}e^{k_jx} \end{aligned} \tag{5}$$

are solutions of (2). It follows that any linear combination of these functions is also a solution, and conversely it can be shown that every solution of (2) is a linear combination of the functions (5) (a proof of this last fact will be given later: see p. 63). Thus

> *the general solution of the reduced equation* (2) *is an arbitrary linear combination of the* n *functions* (5), *where* $k_1, \ldots, k_j$ *are the distinct roots of* (3), *with multiplicities* $r_1, \ldots, r_j$.

When some roots of (3) are complex, they will occur in pairs $\alpha \pm i\omega$ (assuming that the coefficients $a_1, \ldots, a_n$ are real), and one may then replace combinations of $e^{(\alpha+i\omega)x}$ and $e^{(\alpha-i\omega)x}$ by combinations of $e^{\alpha x} \cos \omega x$ and $e^{\alpha x} \sin \omega x$.

Next it can be shown, by an argument similar to that used on p. 10, that the general solution of the full equation (1) is the sum of any one particular solution and of the complementary function (general solution of the reduced equation), the latter involving n arbitrary constants. Particular solutions can be found whenever $f(x)$ is a polynomial, exponential, cosine or sine, by substituting a trial solution of suitable form. The usual troubles occur when $f(x)=Me^{mx}$ and $k=m$ is a root of equation (3), i.e. when e^{mx} already occurs in the complementary function; if $k=m$ is a root of multiplicity r, one must take a trial solution of the form Ax^re^{mx}.

It should hardly be necessary to give a detailed set of instructions (like that on p. 18 for second order equations), but we will illustrate some typical points by an example.

Example 1: *Find the general solution of* $(D^4-1)y=\sin x$.
The roots of $k^2-1=0$ are $k=\pm i, \pm 1$, so that the complementary function is an arbitrary linear combination of $\cos x$, $\sin x$, e^x and e^{-x}. To find a particular solution, replace the right-hand side by e^{ix}

(and take the imaginary part of the resulting particular solution). Since $k=i$ is a root of $k^4-1=0$, we must try $y=Axe^{ix}$. This gives

$$D^4y=A(i^4x+4i^3)e^{ix}=A(x-4i)e^{ix},$$
$$(D^4-1)y=-4iAe^{ix},$$

and we must take $-4iA=1$, $A=\frac{1}{4}i$. The particular solution is

$$I\{\tfrac{1}{4}ixe^{ix}\}=\tfrac{1}{4}x\cos x,$$

and the general solution is

$$y=\tfrac{1}{4}x\cos x+C_1\cos x+C_2\sin x+C_3e^x+C_4e^{-x}.$$

3.2 First order systems. The methods of the present chapter can be adapted to problems involving simultaneous differential equations for several unknowns $y, z, \ldots$, but the arithmetical work tends to become excessively clumsy. The operational method (Ch. II) is more efficient and systematic, and is therefore to be preferred in all except the very simplest problems. Accordingly we content ourselves here with the simplest possible case: a pair of first order equations for two unknowns, y and z,

$$\frac{dy}{dx}+\alpha y+\beta z=0,\ \frac{dz}{dx}+\gamma y+\delta z=0, \tag{6}$$

with zero right-hand sides.

We shall look for solutions of (6) in which y and z are both multiples of the same exponential, say

$$y=Ae^{kx},\ z=Be^{kx}. \tag{7}$$

Naturally we want a solution other than the trivial one $y=z=0$ so A and B should not both be zero. When we substitute into (6), we find after cancelling e^{kx} that

$$(k+\alpha)A+\beta B=0,\ \gamma A+(k+\delta)B=0. \tag{8}$$

From (8) we obtain

$$[(k+\alpha)(k+\delta)-\beta\gamma]A=[(k+\alpha)(k+\delta)-\beta\gamma]B=0$$

so that A and B will both be zero unless k satisfies the quadratic equation

$$(k+\alpha)(k+\delta)-\beta\gamma=0. \tag{9}$$

When k does satisfy (9), then the left-hand members in (8) are proportional and either equation in (8) determines the ratio $A : B$ uniquely. (Those readers who are familiar with determinants will recognize (9), written in the form

$$\begin{vmatrix} k+\alpha & \beta \\ \gamma & k+\delta \end{vmatrix} = 0,$$

as a necessary and sufficient condition for equations (8) to have a non-trivial solution.)

To avoid further complications we will now assume that the quadratic (9) has distinct roots, k_1 and k_2. Taking $k=k_1$ and $k=k_2$ in (8), ratios $A_1 : B_1$ and $A_2 : B_2$ can be found such that

$$y=A_1e^{k_1x},\ z=B_1e^{k_1x} \text{ and } y=A_2e^{k_2x},\ z=B_2e^{k_2x}$$

are solutions of (6). We can then form further solutions by taking arbitrary multiplies of these solutions and adding them together, and it can be shown that every solution of (6) can be obtained in this way.

Example 2:

$$\frac{dy}{dx}-3y+2z=0,\ \frac{dz}{dx}+4y-z=0.$$

The quadratic (9) is $(k-3)(k-1)-2\cdot4=0$, $k^2-4k-5=0$, with roots $k=5$, -1. The equations (8) are

$$(k-3)A+2B=0,\ 4A+(k-1)B=0.$$

When $k=5$, either equation gives $A+B=0$, and when $k=-1$ either equation gives $2A-B=0$. Thus we can take

$$y=e^{5x},\ z=-e^{5x};$$

or

$$y=e^{-x},\ z=2e^{-x};$$

from these we can build up the general solution

$$y=C_1e^{5x}+C_2e^{-x},\ z=-C_1e^{5x}+2C_2e^{-x}.$$

3.3 Arbitrary constants and initial conditions. The general solution to the nth order equation (1), p. 20, contains n arbitrary constants. Their values can be determined if n suitably chosen conditions are imposed on the solution: usually these are initial conditions, and the n functions whose values are prescribed at some given point $x=x_0$ are y and its derivatives up to order $n-1$. We assert (without proof) that *the nth order equation* (1) *has exactly one solution satisfying given initial conditions of the form* $y=y_0$, $dy/dx=y_1, \ldots$, $d^{n-1}y/dx^{n-1}=y_{n-1}$, *when* $x=x_0$.

This assertion will be proved in Ch. II, § 2.6 (though a different notation will there be used).

Similarly the general solution of the pair of equations (6), p. 23, contains two arbitrary constants whose values can be determined when y and z are required to take given values y_0 and z_0 at $x=x_0$:

the pair of first order equations (6) *has exactly one solution satisfying given initial conditions*

$$y=y_0,\ z=z_0, \text{ when } x=x_0.$$

This is a very special case of a general fact about sets of first order equations which will also be established in Ch. II, § 2.6

PROBLEMS FOR CHAPTER I

[y' will denote dy/dx; Dy, D^2y, . . . will denote dy/dx, d^2y/dx^2, . . . Unless otherwise stated, the general solution should be found.]

1. $y'-y=x^2$.

2. $y'-y=e^x$, $y=0$ when $x=0$.

3. $y'+2y=\cos x$.

4. $y'+y=\operatorname{sech} x$.

5. $(D^2+4D+5)y=2e^{-2x}$.

6. $(D^2+1)y=e^{-x}\cos x$.

7. $(D^2-5D+6)y=\cos x+\sin x$.

8. $(D^2-4)y=x^2-3x-4$.

9. $(D^2+2D+1)y=xe^{-x}$.

10. $(D^2+2D+5)y=x\sin x$.

11. $(D^2-2D-3)y=0$; $y=2$, $Dy=-4$, when $x=0$.

12. Find a particular solution of $(D^2+1)y=3\cos x-\sin x$ by substituting the trial solution $y=x(A\cos x+B\sin x)$.

13. $(D^3+1)y=6e^{2x}$.

14. $(D^3-3D^2+D+5)y=100\cos 3x$.

15. $(D^4-2D^2+1)y=x^3$.

16. $Dy=2y-z$, $Dz=3y-2z$; $y=2$, $z=0$, when $x=0$.

17. $Dy+y+z=0$, $Dz-y+z=0$.

CHAPTER TWO

The Operational Method

§ 1. Preliminary discussion of the method

1.1 The operator Q. The purpose of the operational method has already been indicated in Ch. I: it is to find solutions of linear differential equations with constant coefficients which satisfy specified initial conditions.

We shall depart from our previous notations: the unknown will now be called x, and the independent variable t; the initial conditions will be imposed at $t=0$. This notation is in common use because it fits in with many applications in which t is a time variable and x describes the state of some physical system whose initial state (at time 0) is given.

To introduce the operational method we consider the *initial value problem*

$$\left.\begin{aligned} &\frac{dx}{dt}-ax=f(t), \\ &x=x_0 \text{ when } t=0, \end{aligned}\right\} \tag{1}$$

for a first order equation. The solution is given by

$$\frac{d}{dt}(xe^{-at})=f(t)e^{-at},$$

$$x(t)e^{-at}-x(0)=\int_0^t f(\tau)e^{-a\tau}d\tau,$$

$$x(t)=x_0e^{at}+\int_0^t e^{a(t-\tau)}f(\tau)d\tau. \tag{2}$$

Thus (1) can be *solved* by an integration from 0 to t; but it can also be *stated* in a form involving only such integrations, without any differentiations. To see this, integrate each term in

$$\frac{dx}{dt}-ax=f(t)$$

from 0 to t. This gives

$$x(t)-x_0-a\int_0^t x(\tau)d\tau=\int_0^t f(\tau)d\tau. \tag{3}$$

Conversely if x satisfies (3), then $x(0)=x_0$ because the integrals occurring in (3) are zero when $t=0$, and also by differentiating (3) we get

$$\frac{d}{dt}x(t)-ax(t)=f(t).$$

Thus *the initial value problem* (1) *is equivalent to the 'integral equation'* (3). The operational method takes (3) as its starting point, because this single equation incorporates the initial value x_0 whilst in (1) the initial condition had to be stated separately.

The operation 'integrate from 0 to t' will play an essential part from now on, and so we shall use a special notation for it. If f is any function of t, we shall denote by Qf that function whose value at t is

$$Qf(t)=\int_0^t f(\tau)d\tau, \tag{4}$$

and we shall speak of Q as the *operator* which produces Qf when applied to f. In this notation, (3) may be written as

$$x-x_0-aQx=Qf,$$
$$(1-aQ)x=x_0+Qf, \tag{5}$$

where $(1-aQ)x$ naturally stands for $x-aQx$. It is now tempting to write down a solution to (5), namely

$$x=\frac{1}{1-aQ}x_0+\frac{Q}{1-aQ}f, \tag{6}$$

regardless of the fact that $1-aQ$ is not a number but an operator so that the meaning of such expressions as $\dfrac{1}{1-aQ}$ is not clear at present. However, we can tentatively identify the two terms on the right in (6) by comparing with the known solution (2); such a comparison suggests that we should interpret (6) by means of the rules

$$\frac{1}{1-aQ}x_0=x_0e^{at}, \tag{7}$$

$$\frac{Q}{1-aQ}f(t)=\int_0^t e^{a(t-\tau)}f(\tau)d\tau. \tag{8}$$

From (8) we see that $\dfrac{Q}{1-aQ}$ must again be regarded as an operator; so must $\dfrac{1}{1-aQ}$, although this is not so evident because in (7) it operates only on the constant function x_0.

1.2 Formal calculations with Q. So far no advantage has been gained from the introduction of Q, since we already knew the solution to problem (1). Let us now look at an initial value problem for a second order equation:

$$\left.\begin{aligned}&\frac{d^2x}{dt^2}+a\frac{dx}{dt}+bx=f(t),\\ &x=x_0,\ \frac{dx}{dt}=x_1,\ \text{when } t=0.\end{aligned}\right\} \tag{9}$$

We convert this into a form involving integrations only by integrating twice. The first integration gives

$$\frac{dx}{dt}-x_1+a(x-x_0)+bQx=Qf,$$

$$\frac{dx}{dt}+ax+bQx=x_1+ax_0+Qf; \tag{10}$$

the second integration gives

$$x-x_0+aQx+bQ(Qx)=Q(x_1+ax_0)+Q(Qf),$$

$$x+aQx+bQ^2x=x_0+aQx_0+Qx_1+Q^2f, \tag{11}$$

where we have written Q^2f for $Q(Qf)$ so that Q^2 denotes repeated integration. Now suppose conversely that x satisfies (11). Then $x=x_0$ when $t=0$, because all terms in (11) which involve Q or Q^2 are zero when $t=0$. Next, by differentiating (11) we get back to (10), and on putting $t=0$ in (10) we find that $\dfrac{dx}{dt}=x_1$ when $t=0$. Finally, by differentiating (10) we get back to (9). Thus we have again converted the initial value problem

(9) into an equivalent single equation, (11), which involves integrations only and incorporates the initial values x_0 and x_1.

We now write the left-hand side of (11) as

$$(1+aQ+bQ^2)x$$

and factorize this as

$$(1-\lambda Q)(1-\mu Q)x,$$

where λ and μ are the roots of $k^2+ak+b=0$; this brings (11) to the form

$$(1-\lambda Q)(1-\mu Q)x=x_0+aQx_0+Qx_1+Q^2f.$$

Equations of this kind can be solved quite easily, provided that we are prepared to manipulate expressions involving Q rather freely; the following example illustrates what kind of manipulations are likely to occur.

Example 1:

$$\left.\begin{aligned} \frac{d^2x}{dt^2}-3\frac{dx}{dt}+2x=f(t), \\ x=1,\ \frac{dx}{dt}=-1, \text{ when } t=0. \end{aligned}\right\}$$

By integrating twice we get (as in (11)) the equivalent equation

$$(1-3Q+2Q^2)x=1-3Q(1)+Q(-1)+Q^2f,$$
$$(1-Q)(1-2Q)x=(1-4Q)1+Q^2f.$$

Proceeding quite formally, we get

$$x=\frac{1-4Q}{(1-Q)(1-2Q)}1+\frac{Q^2}{(1-Q)(1-2Q)}f.$$

To deal with the first term on the right, we note that we can evaluate $\frac{1}{1-Q}1$ and $\frac{1}{1-2Q}1$ by using formula (7), p. 27. We shall therefore evaluate $\frac{1-4Q}{(1-Q)(1-2Q)}1$ by putting it into partial fractions; thus

$$\frac{1-4Q}{(1-Q)(1-2Q)}1=\left(\frac{3}{1-Q}-\frac{2}{1-2Q}\right)1=3e^t-2e^{2t}.$$

We can deal similarly with $\frac{Q^2}{(1-Q)(1-2Q)}f$ by writing it as

$$\frac{Q}{(1-Q)(1-2Q)}Qf=\left(\frac{-1}{1-Q}+\frac{1}{1-2Q}\right)Qf=-\frac{Q}{1-Q}f+\frac{Q}{1-2Q}f;$$

the last expression can now be evaluated from formula (8), p. 28. So we arrive at the tentative solution

$$x=3e^t-2e^{2t}+\int_0^t(e^{2(t-\tau)}-e^{(t-\tau)})f(\tau)d\tau.$$

Some readers may find it interesting to verify that this is a solution; to do this, they may prefer to write the last term on the right as

$$e^{2t}\int_0^t e^{-2\tau}f(\tau)d\tau-e^t\int_0^t e^{-\tau}f(\tau)d\tau.$$

Naturally we must still show that the treatment to which we subjected Q in the above example is legitimate. This will be done in the next few sections. However, some readers may be eager to learn the technique of the operational method as quickly as possible, and may be prepared to accept formal calculations with Q at face value. Such readers may, if they wish, pass straight on to § 1.6 in which the technique will be further developed.

1.3 Operators. Whenever we have a rule for producing from any given function f a new function Af, we shall call A an *operator*. For example, the rules

$$Qf(t)=\int_0^t f(\tau)d\tau,$$

$$Df(t)=f'(t),$$

define the operators Q (integration) and D (differentiation). By 'function' we shall always mean a function of t which possesses derivatives of every order, i.e. which can be differentiated as many times as we please.

We call A a *linear* operator if

$$A(\lambda f+\mu g)=\lambda Af+\mu Ag,$$

for any functions f, g and constants λ, μ; here $A(\lambda f+\mu g)$ stands for the result of applying the operator A to the function $\lambda f+\mu g$. Clearly Q and D, as defined above, are linear.

If A and B are linear operators, they may be combined in two ways. First, we may take two constants α and β, and form the linear combination $\alpha A+\beta B$. The effect of this operator on any function f is defined by

$$(\alpha A+\beta B)f=\alpha Af+\beta Bf.$$

Secondly we may form the *product* AB. The effect of this operator on f is defined by

$$ABf=A(Bf).$$

Thus:

to find ABf, first apply B to f, and then apply A to the result Bf.

It is almost obvious that $\alpha A+\beta B$ and AB are linear operators when A and B are linear, and we shall omit the formal verification of this fact.

The definition of product will be illustrated by calculating QD and DQ. To find QDf, first apply D to f (i.e. differentiate) and then apply Q (i.e. integrate from 0 to t):

$$QDf(t)=\int_0^t f'(\tau)d\tau=f(t)-f(0). \tag{12}$$

Similarly

$$DQf(t)=\frac{d}{dt}\int_0^t f(\tau)d\tau=f(t). \tag{13}$$

Notice that QD and DQ are *not* identical, and we must therefore take care to attend to the order of the factors when dealing with products of operators. When AB and BA *are* equal, we shall say that A and B *commute*.

When one performs numerical calculations, one uses the laws of algebra quite freely. We will not give an exhaustive list of these laws, but recall several typical ones:

$a.b=b.a$ ('commutative law of multiplication'),
$a.(b+c)=a.b+a.c$ ('distributive law'),
$a.(b.c)=(a.b).c$ ('associative law of multiplication').

Analogous laws also hold for calculations with operators, with one important exception (already noted): the commutative law $AB=BA$ does not always hold. All other laws relating to addition and multiplication continue to hold; for instance the distributive law, $A(B+C)=AB+AC$, and the associative law, $A(BC)=(AB)C$ where A, B and C are linear operators. The associative law means that we can simply write ABC, without any brackets, for the operator whose effect on f is found by first applying C to f, then applying B to the result Cf, and finally applying A to the result $B(Cf)$.

In particular we may form the *powers* of a single operator A:

$$A^2=AA,\ A^3=AAA,\ \ldots,$$

so that A^n stands for 'A applied n times in succession'. We can then form *polynomials* in A:

$$c_0+c_1A+c_2A^2+\ \ldots\ +c_nA^n,$$

where c_0 denotes the operation of multiplying by c_0. Any two such polynomials commute, so that polynomials in A may be multiplied together or factorized just like ordinary polynomials, the order of the factors being immaterial. For instance

$$(1+A)(1-2A)=(1-2A)(1+A)=1-A-2A^2.$$

1.4 The inverse of an operator. Now consider an equation of the form

$$Ax=g,$$

where A is a linear operator, g is a given function and x is an unknown function. (Equations of this kind have already occurred several times, for instance (5) on p. 27 and (11) on p. 28.) Such an equation may have no solutions at all, or on the other hand may have infinitely many solutions. For example $Qx=g$, that is

$$\int_0^t x(\tau)d\tau=g(t),$$

has no solution if $g(0)\neq 0$ because the left-hand side is necessarily zero when $t=0$; but $Dx=g$, that is

$$\frac{dx}{dt}=g(t),$$

always has infinitely many solutions:

$$x(t)=\int_0^t g(\tau)d\tau+C.$$

However, *suppose that $Ax=g$ has exactly one solution for each given g*. This solution will of course depend on g; *we shall denote it by $A^{-1}g$*, and call the operator A^{-1} the *inverse* of A. The operator A^{-1} is again linear, i.e. we have

$$A^{-1}(\lambda g_1+\mu g_2)=\lambda A^{-1}g_1+\mu A^{-1}g_2.$$

To prove this, write $x_1=A^{-1}g_1$ and $x_2=A^{-1}g_2$, so that $Ax_1=g_1$ and

$Ax_2 = g_2$. Then

$$A(\lambda x_1 + \mu x_2) = \lambda A x_1 + \mu A x_2 = \lambda g_1 + \mu g_2,$$

and therefore $x = \lambda x_1 + \mu x_2$ is *a* solution of the equation

$$Ax = \lambda g_1 + \mu g_2.$$

But this equation has only one solution, namely

$$x = A^{-1}(\lambda g_1 + \mu g_2),$$

and so $\lambda x_1 + \mu x_2$ must coincide with the above expression; in other words we have

$$\lambda A^{-1} g_1 + \mu A^{-1} g_2 = A^{-1}(\lambda g_1 + \mu g_2),$$

which proves our assertion that A^{-1} is linear. Next, we have

$$A(A^{-1}g) = g;$$

this merely states the fact that $x = A^{-1}g$ is a solution of $Ax = g$. But we also have

$$A^{-1}(Ag) = g,$$

because $x = A^{-1}(Ag)$ is by definition the solution of $Ax = Ag$; now $x = g$ is clearly *a* solution of this equation and since the equation has only one solution, g must coincide with $A^{-1}(Ag)$. We have now evaluated the product of A and A^{-1}, in either order, and our results can be written as

$$AA^{-1} = \mathrm{I}, \ A^{-1}A = \mathrm{I},$$

where I denotes the 'identity operator' defined by $\mathrm{I}g = g$. Conversely, suppose that we can find a linear operator B such that

$$AB = \mathrm{I}, \ BA = \mathrm{I}.$$

Then B is the inverse of A; that is, the equation $Ax = g$ has exactly one solution, given by $x = Bg$. To see this, note that $x = Bg$ *is* a solution because

$$A(Bg) = ABg = \mathrm{I}g = g,$$

and that it is the only solution because if $Ax = g$ then

$$x = \mathrm{I}x = BAx = B(Ax) = Bg.$$

To sum up the preceding discussion:

If the equation $Ax = g$ has exactly one solution, $x = A^{-1}g$, for each given g, then the operator A^{-1} is linear, and $AA^{-1} = A^{-1}A = \mathrm{I}$; A^{-1} will be called the inverse of A.

Conversely if there exists a linear operator B such that $AB = BA = \mathrm{I}$, then A has an inverse, namely $A^{-1} = B$.

To illustrate the definition of the inverse operator, let us first observe that neither Q nor D possesses an inverse; the equation

$Qx=g$ sometimes has no solution, whilst $Dx=g$ always has infinitely many solutions. Next, we will show that $1-aQ$ (where a is any constant) does have an inverse; this important fact forms the basis of the operational method. To establish it, we must look at the equation

$$(1-aQ)x=g,$$

that is

$$x(t)-a\int_0^t x(\tau)d\tau=g(t).$$

This is equivalent to

$$x'(t)-ax(t)=g'(t),\ x(0)=g(0),$$

and we know from Ch. I (§ 1.2) that this initial value problem has exactly one solution:

$$x(t)=g(0)e^{at}+\int_0^t e^{a(t-\tau)}g'(\tau)d\tau. \tag{14}$$

Thus $(1-aQ)^{-1}$ exists, and $x=(1-aQ)^{-1}g$ can be calculated from formula (14). (At this point, we should recall the agreement made on p. 30 that all functions are to have derivatives of every order. Now $x(t)$, given by (14), is such a function; its first derivative exists and equals $ax(t)+g'(t)$, its second derivative therefore exists and equals $ax'(t)+g''(t)$, and so on.)

In particular if we take $g(t)=1$, identically, we have $g(0)=1$ and $g'(\tau)=0$; if we take $g=Qf$, then $g(0)=0$ and $g'(\tau)=f(\tau)$; from (14) we therefore obtain the formulae

$$(1-aQ)^{-1}1=e^{at}, \tag{15}$$

$$(1-aQ)^{-1}Qf(t)=\int_0^t e^{a(t-\tau)}f(\tau)d\tau. \tag{16}$$

These are precisely the formulae (7) and (8) which we tentatively wrote down in § 1.1, except that there we wrote $\dfrac{1}{1-aQ}$ and $\dfrac{Q}{1-aQ}$ instead of $(1-aQ)^{-1}$ and $(1-aQ)^{-1}Q$.

1.5 Inverse of a product. When we come to deal with second order equations, as in § 1.2, we are led to equations of the form

$$(1+aQ+bQ^2)x=(c_1+c_2Q)1+Q^2f.$$

It will then be useful to know the inverse of $1+aQ+bQ^2$; we shall see presently that this operator *has* an inverse because it can be factorized as $(1-\lambda Q)(1-\mu Q)$ and we already know the inverses of the factors $1-\lambda Q$ and $1-\mu Q$.

Consider then the following more general question; given that two operators A, B have inverses, does AB also have an inverse? To answer this question, we look at the equation

$$ABx=g.$$

This can be solved in two steps; first write the equation as

$$A(Bx)=g,$$

and solve for Bx. This gives

$$Bx=A^{-1}g,$$

and now we can solve for x, obtaining

$$x=B^{-1}(A^{-1}g)=B^{-1}A^{-1}g.$$

This is the only solution, and therefore AB does have an inverse, given by

$$(AB)^{-1}=B^{-1}A^{-1}.$$

We can also verify this by showing that the product of AB and $B^{-1}A^{-1}$, in either order, is 1. Thus

$$(AB)(B^{-1}A^{-1})=A(BB^{-1})A^{-1}=A(1)A^{-1}=AA^{-1}=1,$$
$$(B^{-1}A^{-1})(AB)=B^{-1}(A^{-1}A)B=B^{-1}(1)B=B^{-1}B=1.$$

Now suppose in addition that A and B commute. Then $(BA)^{-1}=A^{-1}B^{-1}$, but $BA=AB$ so that we can now write

$$(AB)^{-1}=A^{-1}B^{-1}=B^{-1}A^{-1}.$$

Thus:

I { *if A and B commute and have inverses A^{-1} and B^{-1}, then AB has inverse $B^{-1}A^{-1}=A^{-1}B^{-1}$, and so A^{-1} and B^{-1} also commute.*

In particular, if A has an inverse then $A^2=AA$ has inverse $A^{-1}A^{-1}=(A^{-1})^2$, $A^3=A^2A$ has inverse $(A^2)^{-1}A^{-1}=(A^{-1})^2A^{-1}$ $=(A^{-1})^3$, and generally

$$(A^n)^{-1}=(A^{-1})^n.$$

We shall therefore denote the inverse of A^n by A^{-n}, and we then have $A^{-n}A^n=A^nA^{-n}=1$. Indeed we have now defined A^r for positive and negative integers r; if we define $A^0=1$, then the 'index law' $A^rA^s=A^{r+s}$ will hold for all r and s (positive, negative or zero).

The following fact will also be useful:

II { *if A has an inverse, and if A commutes with B, then A^{-1} also commutes with B; that is, $A^{-1}B=BA^{-1}$.*

To prove this, take the equation $AB = BA$ and multiply by A^{-1} at the beginning and end:

$$A^{-1}(AB)A^{-1} = A^{-1}(BA)A^{-1},$$
$$(A^{-1}A)BA^{-1} = A^{-1}B(AA^{-1});$$

since $A^{-1}A = AA^{-1} = 1$, the last equation reduces to $BA^{-1} = A^{-1}B$ as required.

The preceding work may seem remote from the practical problem of solving differential equations, but in fact it will be very useful to us because we shall be dealing with products of operators in which the factors are all of the form $1-\lambda Q$, $1-\mu Q$, . . , and so commute with each other, and also commute with Q. We shall therefore be able to use rules I and II above quite often, and by doing this we shall be able to handle calculations with Q just like ordinary (numerical) calculations. For example, II shows that

$$(1-aQ)^{-1}Q = Q(1-aQ)^{-1},$$

and I shows that if $1 + aQ + bQ^2 = (1-\lambda Q)(1-\mu Q)$, the inverse of $1 + aQ + bQ^2$ exists and is given by

$$(1 + aQ + bQ^2)^{-1} = (1-\lambda Q)^{-1}(1-\mu Q)^{-1} = (1-\mu Q)^{-1}(1-\lambda Q)^{-1}.$$

Again, if we have an expression like

$$(1-Q)^{-1}(1-3Q)^{-1}(1-Q)^{-1}(1-3Q)$$

we can 'cancel the common factor $1-3Q$' by using I to reverse the order of the two middle factors, which gives

$$(1-Q)^{-1}(1-Q)^{-1}(1-3Q)^{-1}(1-3Q) = ((1-Q)^{-1})^2 = (1-Q)^{-2}.$$

We can now afford to simplify our notation, without any risk of ambiguity, by writing

$$\frac{1}{1-aQ} \quad \text{for } (1-aQ)^{-1},$$

$$\frac{Q}{1-aQ} \quad \text{for } (1-aQ)^{-1}Q = Q(1-aQ)^{-1},$$

$$\frac{Q}{(1-aQ)^2} \quad \text{for } (1-aQ)^{-2}Q = Q(1-aQ)^{-2},$$

and so on; more generally if we have any expression

$$\frac{\phi(Q)}{(1-a_1 Q) \ . \ . \ . \ (1-a_n Q)}$$

where $\phi(Q)$ is a polynomial in Q, it shall have the interpretation

$$(1-a_1Q)^{-1} \ . \ . \ . \ (1-a_nQ)^{-1}\phi(Q)$$

and we may then alter the order of the factors in any way we please. If it so happens that $\phi(Q)$ contains say $(1-a_2Q)$ as a factor, we can cancel this factor against $(1-a_2Q)^{-1}$ by a suitable re-ordering which brings $(1-a_2Q)^{-1}$ and $(1-a_2Q)$ next to each other.

It is now easy to justify 'partial fraction expansions' such as those which we used tentatively in Example 1 (p. 29). Thus, to put $\frac{1-4Q}{(1-Q)(1-2Q)}$ into partial fractions, write the numerator as

$$1-4Q=3(1-2Q)-2(1-Q);$$

then

$$\frac{1-4Q}{(1-Q)(1-2Q)}=\frac{3(1-2Q)}{(1-Q)(1-2Q)}-\frac{2(1-Q)}{(1-Q)(1-2Q)}=\frac{3}{1-Q}-\frac{2}{1-2Q},$$

on cancelling the appropriate common factors. Similar justifications can be given in more complicated examples; thus any rational function of Q, of the form

$$\frac{c_0+c_1Q+\dots+c_{n-1}Q^{n-1}}{(1-a_1Q)\dots(1-a_nQ)}$$

may be put into partial fractions of the form

$$\frac{A_1}{1-a_1Q}+\frac{A_2}{1-a_2Q}+\dots+\frac{A_n}{1-a_nQ}$$

just as if Q were an ordinary (numerical) variable. (Of course the form of the expansion must be modified when the denominator $(1-a_1Q)\dots(1-a_nQ)$ contains repeated factors; this point will be dealt with in the next section.)

1.6 Partial fractions for inverses. We have now reached a point where we can justify any of the formal calculations with Q which we shall perform, and it is time to develop the technique of such calculations somewhat further. We already know how to evaluate $\frac{1}{1-aQ}1$ and $\frac{Q}{1-aQ}f$, but we shall also need the more general expressions

$$\frac{Q^n}{(1-aQ)^{n+1}}1,\ \frac{Q^{n+1}}{(1-aQ)^{n+1}}f \quad (n=1, 2, 3, \dots),$$

which we shall now evaluate.

First, using (7) and (8) from pp. 27–8, we have

$$\frac{Q}{(1-aQ)^2}1=\frac{Q}{1-aQ}\left(\frac{1}{1-aQ}1\right)=\frac{Q}{1-aQ}(e^{at})$$
$$=\int_0^t e^{a(t-\tau)}e^{a\tau}d\tau$$
$$=e^{at}\int_0^t 1d\tau=te^{at}.$$

Clearly this argument can be repeated:

$$\frac{Q^2}{(1-aQ)^3}1=\frac{Q}{1-aQ}\left(\frac{Q}{(1-aQ)^2}1\right)=\frac{Q}{1-aQ}(te^{at})$$
$$=\int_0^t e^{a(t-\tau)}\tau e^{a\tau}d\tau=e^{at}\int_0^t \tau d\tau=\frac{t^2}{2}e^{at};$$
$$\frac{Q^3}{(1-aQ)^4}1=\frac{Q}{1-aQ}\left(\frac{Q^2}{(1-aQ)^3}1\right)=\frac{Q}{1-aQ}\left(\frac{t^2}{2}e^{at}\right)$$
$$=e^{at}\int_0^t \frac{\tau^2}{2}d\tau=\frac{t^3}{3!}e^{at}.$$

Proceeding in this way, we get the general result

$$\frac{Q^n}{(1-aQ)^{n+1}}1=\frac{t^n}{n!}e^{at} \qquad (17)$$

for $n=1, 2, 3, \ldots$; it also holds for $n=0$ when we make the usual definitions that $Q^0=1$, $t^0=1$ and $0!=1$.

Notice that we have chosen to evaluate $\frac{Q}{(1-aQ)^2}1$ rather than $\frac{1}{(1-aQ)^2}1$. The latter is given by

$$\frac{1}{(1-aQ)^2}1=\frac{(1-aQ)+aQ}{(1-aQ)^2}1=\frac{1}{1-aQ}1+a\frac{Q}{(1-aQ)^2}1$$
$$=e^{at}+ate^{at}=(1+at)e^{at}$$

and is thus less simple than $\frac{Q}{(1-aQ)^2}1$.

The second formula to be established reads

$$\frac{Q^{n+1}}{(1-aQ)^{n+1}}f(t)=\int_0^t \frac{(t-\tau)^n}{n!}e^{a(t-\tau)}f(\tau)d\tau. \qquad (18)$$

This holds when $n=0$, and we shall prove it by induction for $n=1, 2, 3, \ldots$. Suppose that (18) has been established for $n=k$. Then

$$\frac{Q^{k+2}}{(1-aQ)^{k+2}}f = \frac{Q}{1-aQ}\left\{\frac{Q^{k+1}}{(1-aQ)^{k+1}}f\right\},$$

$$\frac{Q^{k+2}}{(1-aQ)^{k+2}}f(t) = \int_0^t e^{a(t-\tau)}\left\{\frac{Q^{k+1}}{(1-aQ)^{k+1}}(\tau)\right\}d\tau$$

$$= \int_0^t e^{a(t-\tau)}\left\{\int_0^\tau \frac{(\tau-u)^k}{k!}e^{a(\tau-u)}f(u)du\right\}d\tau.$$

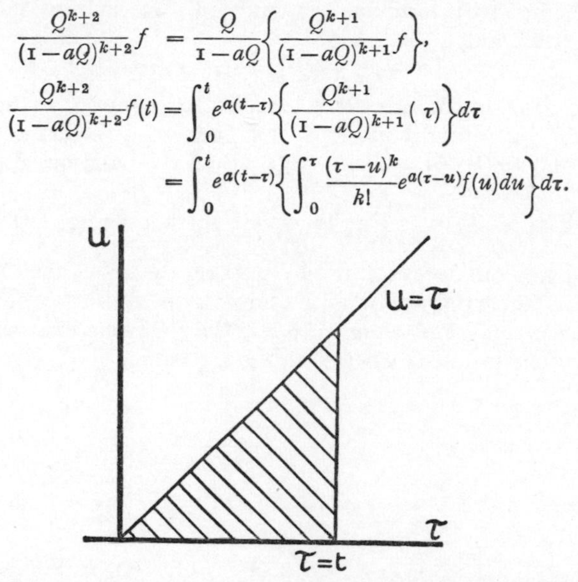

In this repeated integral we can invert the order of the integrations, provided that we attend carefully to the limits of integration. We have $0 \leqslant u \leqslant \tau \leqslant t$, so if we do the τ-integration first (keeping u fixed), then τ varies from u to t; in the second integration, u varies from 0 to t. (This can also be seen by regarding the integral as a *double* integral, taken over the shaded triangle in the (τ, u)-plane in the diagram. If the u-integration is done first, u goes from 0 to τ; if the τ-integration is done first, τ goes from u to t; in either case the second integration is from 0 to t.) Thus we get

$$\int_0^t\left\{\int_u e^{a(t-\tau)}\frac{(\tau-u)^k}{k!}e^{a(\tau-u)}f(u)d\tau\right\}du$$

$$= \int_0^t e^{a(t-u)}f(u)\left\{\int_u^t \frac{(\tau-u)^k}{k!}d\tau\right\}du$$

$$= \int_0 e^{a(t-u)}f(u)\frac{(t-u)^{k+1}}{(k+1)!}du.$$

This is (18) for $n=k+1$, except that the integration variable is called u instead of τ, and the induction has been completed.

Armed with formulae (17) and (18), we can now tackle any equation like

$$(1+aQ+bQ^2)x=(c_1+c_2Q)1+Q^2f, \tag{19}$$

and also similar but more complicated equations which arise from differential equations of higher than second order. Let $(1+aQ+bQ^2)=(1-\lambda Q)(1-\mu Q)$; then the solution to (19) is

$$x=\frac{c_1+c_2Q}{(1-\lambda Q)(1-\mu Q)}1+\frac{Q^2}{(1-\lambda Q)(1-\mu Q)}f,$$

and this can be evaluated by putting the right-hand side into suitable partial fractions. Example 1 (p. 29) shows how this is done when λ and μ are unequal; the following example shows how one proceeds when λ and μ are equal.

Example 2:

Evaluate $$\frac{2-Q}{1-2Q+Q^2}1+\frac{Q^2}{1-2Q+Q^2}f.$$

Here $1-2Q+Q^2=(1-Q)^2$. We could express $\frac{2-Q}{(1-Q)^2}$ in terms of of $\frac{1}{1-Q}$ and $\frac{1}{(1-Q)^2}$, but it is more convenient to use $\frac{Q}{(1-Q)^2}$ rather than $\frac{1}{(1-Q)^2}$. So we write $2-Q$ as $2(1-Q)+Q$, and then we have

$$\frac{2-Q}{(1-Q)^2}1=\frac{2(1-Q)+Q}{(1-Q)^2}1=\frac{2}{1-Q}1+\frac{Q}{(1-Q)^2}1$$
$$=2e^t+te^t, \text{ by (17)}.$$

The second term can be evaluated at once from (18), with $n=1$, and we therefore have

$$\frac{2-Q}{(1-Q)^2}1+\frac{Q^2}{(1-Q)^2}f(t)=(2+t)e^t+\int_0^t(t-\tau)e^{(t-\tau)}f(\tau)d\tau.$$

Similar principles apply to the evaluation of the more general expressions

$$\frac{\phi(Q)}{1+a_1Q+\ldots+a_nQ^n}1,\ \frac{Q^n}{1+a_1Q+\ldots+a_nQ^n}f, \tag{20}$$

where $\phi(Q)$ is a polynomial of degree less than n. Factorize $1+a_1Q+\ldots+a_nQ^n$ as $(1-\lambda_1Q)^{m_1}\ldots(1-\lambda_kQ)^{m_k}$, and then express

$$\frac{\phi(Q)}{(1-\lambda_1Q)^{m_1}\ldots(1-\lambda_kQ)^{m_k}}$$

as a combination of

$$\frac{1}{1-\lambda_1Q}, \frac{Q}{(1-\lambda_1Q)^2}, \ldots, \frac{Q^{m_1-1}}{(1-\lambda_1Q)^{m_1}}$$

and similar terms involving $1-\lambda_2Q, \ldots, 1-\lambda_kQ$; likewise write $\dfrac{Q^n}{1+\ldots+a_nQ^n}$ as a combination of

$$\frac{Q}{1-\lambda_1Q}, \ldots, \frac{Q^{m_1}}{(1-\lambda_1Q)^{m_1}}$$

and similar terms. The expressions (20) can then be evaluated by using formulae (17) and (18).

In general the use of formula (18) will lead to integrals which cannot be further simplified. However, in practice the function f (which stands on the right-hand side of the differential equation) is often a simple exponential, or polynomial, or a product of a polynomial and an exponential. The solution can then be written in a form which contains no integrals at all; the following example shows how and why this can be done.

Example 3:

Find the solution of $\dfrac{d^2x}{dt^2}-\dfrac{dx}{dt}-2x=3e^{-t}$

for which $x=2$ *and* $\dfrac{dx}{dt}=0$ *when* $t=0$.

The equivalent equation involving Q is found as on p. 28; it is

$$(1-Q-2Q^2)x=(2-2Q)1+Q^2(3e^{-t}).$$

We now use (17) to replace $3e^{-t}$ by $\dfrac{3}{1+Q}1$; this gives

$$(1-Q-2Q^2)x=(2-2Q)1+\frac{3Q^2}{1+Q}1,$$

$$(1+Q)(1-2Q)x=\frac{2+Q^2}{1+Q}1,$$

$$x=\frac{2+Q^2}{(1+Q)^2(1-2Q)}1.$$

Now write $\frac{2+Q^2}{(1+Q)^2(1-2Q)}$ as a combination of $\frac{1}{1+Q}$, $\frac{Q}{(1+Q)^2}$, $\frac{1}{1-2Q}$:

$$x=\left(\frac{1}{1+Q}-\frac{Q}{(1+Q)^2}+\frac{1}{1-2Q}\right)1$$

and finally evaluate this by using (17). The solution is

$$x=e^{-t}-te^{-t}+e^{2t}.$$

The above procedure can clearly be applied whenever $f(t)$ is a multiple of $t^n e^{at}$ or a linear combination of such multiples. Only formula (17) is needed, and the result will be of the same general form as $f(t)$ (but with some extra exponential terms which belong to the complementary function).

§ 2. Practical instructions for using the method

2.1 The symbol *p*. The general principles underlying the operational method have now been established. For practical purposes, it is convenient to make a change of notation which will help to simplify the algebraic work involved in handling rational functions of Q. We introduce a new symbol, p, make the substitution $Q=p^{-1}$, and then perform all our calculations with p, treating expressions in p according to the usual algebraic laws.

In this notation, $\frac{Q^m}{(1-\lambda Q)^{m+1}}$ will be replaced by $\frac{p^{-m}}{(1-\lambda p^{-1})^{m+1}}$, which simplifies to $\frac{p}{(p-\lambda)^{m+1}}$. Thus the basic formula (17) from p. 38 becomes

$$\frac{p}{(p-\lambda)^{m+1}}1=\frac{t^m}{m!}e^{\lambda t}, \tag{1}$$

and similarly (18) becomes

$$\frac{1}{(p-\lambda)^{m+1}}f(t)=\int_0^t \frac{(t-\tau)^m}{m!}e^{\lambda(t-\tau)}f(\tau)d\tau. \tag{2}$$

The special cases

$$\left.\begin{aligned} &\frac{1}{p}1=t,\ \frac{1}{p^2}1=\tfrac{1}{2}t^2;\\ &\frac{p}{p-\lambda}1=e^{\lambda t},\ \frac{p}{(p-\lambda)^2}1=te^{\lambda t};\\ &\frac{1}{p}f(t)=\int_0^t f(\tau)d\tau,\ \frac{1}{p-\lambda}f(t)=\int_0^t e^{\lambda(t-\tau)}f(\tau)d\tau \end{aligned}\right\} \tag{3}$$

should be particularly noted.

The typical expression

$$\frac{c_0+c_1Q+\ldots+c_{n-1}Q^{n-1}}{1+a_1Q+\ldots+a_nQ^n}1+\frac{Q^n}{1+\ldots+a_nQ^n}f(t)$$

now becomes

$$p\frac{c_0p^{n-1}+c_1p^{n-2}+\ldots+c_{n-1}}{p^n+a_1p^{n-1}+\ldots+a_n}1+\frac{1}{p^n+\ldots+a_n}f(t)$$

$$=p\frac{C(p)}{A(p)}1+\frac{1}{A(p)}f(t), \text{ say.}$$

The advantage of the latter form is that we can put $C(p)/A(p)$ and $1/A(p)$ into partial fractions of the usual kind, containing terms in

$$\frac{1}{p-\lambda},\ \frac{1}{(p-\lambda)^2},\ \ldots,\ \frac{1}{(p-\lambda)^m}$$

for each factor $(p-\lambda)^m$ of $A(p)$, and then

$$p\frac{C(p)}{A(p)}1+\frac{1}{A(p)}f(t)$$

will contain terms in

$$\frac{p}{p-\lambda}1,\ \ldots,\ \frac{p}{(p-\lambda)^m}1,\ \frac{1}{p-\lambda}f(t),\ \ldots,\ \frac{1}{(p-\lambda)^m}f(t),$$

which can be evaluated from formulae (1) and (2). Standard methods for finding partial fractions are therefore available; if we were to use Q we should have to modify these methods because we should need fractions of the less convenient form

$$\frac{1}{1-\lambda Q},\ \frac{Q}{(1-\lambda Q)^2},\ \ldots,\ \frac{Q^{m-1}}{(1-\lambda Q)^m}.$$

It should be stressed that the symbol p merely serves to simplify calculations, and that it is *not* an operator; the operator on which the method depends is $p^{-1}=Q$. (If p were an operator, it ought to be the inverse of Q; but, as we saw on p. 33, Q does not have an inverse.) The justification for using p lies in the fact that any valid calculation with Q can be 'translated' into an equivalent calculation with p, and vice versa; for instance the p-calculation

$$p\frac{2p+1}{p^2-1}=p\frac{\frac{3}{2}(p+1)+\frac{1}{2}(p-1)}{p^2-1}=\frac{3}{2}\frac{p}{p-1}+\frac{1}{2}\frac{p}{p+1}$$

is equivalent to the Q-calculation

$$\frac{2+Q}{1-Q^2}=\frac{\frac{3}{2}(1+Q)+\frac{1}{2}(1-Q)}{1-Q^2}=\frac{3}{2}\frac{1}{1-Q}+\frac{1}{2}\frac{1}{1+Q},$$

and we already know that such calculations with Q are permitted.

2.2 Procedure for solving nth order equations. To deal with the initial value problem

$$\left.\begin{aligned}(D^n+a_1D^{n-1}+\ldots+a_n)x=f(t),\\ x=x_0,\ Dx=x_1,\ \ldots,\ D^{n-1}x=x_{n-1}\text{ when }t=0,\end{aligned}\right\}\qquad(4)$$

we integrate n times. This leads to an equivalent equation, in terms of Q, into which the initial values $x_0, \ldots, x_{n-1}$ have been incorporated. If we then put $Q=p^{-1}$ and multiply by p^n in order to remove negative powers of p, we shall arrive at the following equation:

$$\begin{aligned}&(p^n+a_1p^{n-1}+\ldots+a_n)x\\ &=f(t)+\left\{\begin{aligned}x_0(p^n+a_1p^{n-1}+\ldots+a_{n-1}p)+x_1(p^{n-1}\\ +a_1p^{n-2}+\ldots+a_{n-2}p)+\ldots+x_{n-1}p\end{aligned}\right\}1.\end{aligned}\qquad(5)$$

This will be called the *operational form* of (4). The details of the calculation which leads to it will be given later (p. 62); we can check it for $n=2$ by putting $Q=p^{-1}$ in (11), p. 28, and multiplying by p^2. (Note that a and b in (11) are now called a_1 and a_2.) At present we want to show how (5) can most easily be memorized, and we recommend the following *rule for finding the operational form* of (4):

On the left-hand side of the differential equation, replace D by the symbol p;
on the right-hand side, add to $f(t)$ the polynomial in p which is obtained from

$$(p^n+a_1p^{n-1}+\ldots+a_n)\left(x_0+\frac{x_1}{p}+\ldots+\frac{x_{n-1}}{p^{n-1}}\right)1$$

by multiplying out and retaining only the positive powers of p.

It should be noted that the polynomial on the right of (5) always contains p as a factor, and also that none of its terms contain the coefficient a_n.

The function $f(t)$ is often the sum of one or more terms of the type $At^r e^{\alpha t}$. Using (1), we can then write $f(t)$ as a sum of terms $Ar!\dfrac{p}{(p-\alpha)^{r+1}}1$; the resulting expression, when simplified, will have the form $F(p)1$ and will be called the *operational form* of $f(t)$. For example, $t^2+(3t-2)e^{-t}$ has the operational form

$$\frac{2}{p^2}1+3\frac{p}{(p+1)^2}1-2\frac{p}{p+1}1=\left(\frac{2}{p^2}+\frac{p(1-2p)}{(p-1)^2}\right)1.$$

When we replace $f(t)$ by $F(p)1$ in equation (5), we obtain

$$A(p)x=(pC(p)+F(p))1.$$

If we solve for x, writing the solution in the form

$$x=p\frac{G(p)}{H(p)}1,$$

we can then put the rational function $G(p)/H(p)$ into partial fractions and evaluate x by using formula (1). It is important to remember that a factor p must be extracted before finding the partial fractions, because we want to express x as a combination of terms $\dfrac{p}{(p-\lambda)^m}1$ $\left(\text{and not } \dfrac{1}{(p-\lambda)^m}1\right)$. This point is accounted for in the following *rule for solving* (4) *when the operational form of $f(t)$ is known*:

Write down the operational form (5) *of* (4), *and at the same time replace $f(t)$ by its operational form $F(p)1$. Solve for $\dfrac{x}{p}$, obtaining* *it in the form* $\frac{x}{p}=R(p)1$ *where* $R(p)$ *is a rational function. Put* $R(p)$ *into partial fractions, and then evaluate* $x=pR(p)1$ *from formula* (1).

Let us now illustrate the procedure by two examples.

Example 1: *Solve* $(D^3+2D^2-D-2)x=3e^{2t}$,

for the initial conditions $x_0=2,\ x_1=-3,\ x_2=1$.

The operational form of $3e^{2t}$ is $\frac{3p}{p-2}1$; the positive powers of p in $(p^3+2p^2-p)\left(2-\frac{3}{p}+\frac{1}{p^2}\right)$ are $2p^3+p^2-7p$. Hence the operational form of the problem is

$$(p^3+2p^2-p-2)x=\left(2p^3+p^2-7p+\frac{3p}{p-2}\right)1$$

$$=\frac{2p^4-3p^3-9p^2+17p}{p-2}1.$$

Factorizing p^3+2p^2-p-2 as $(p-1)(p+1)(p+2)$, we get

$$\frac{x}{p}=\frac{2p^3-3p^2-9p+17}{(p-1)(p+1)(p+2)(p-2)}1$$

$$=\left(-\frac{7/6}{p-1}+\frac{7/2}{p+1}-\frac{7/12}{p+2}+\frac{1/4}{p-2}\right)1.$$

(We have suppressed the calculations required for the partial fraction expansion; the technique of such calculations will be discussed in the next section.) Thus we have

$$x=-\frac{7}{6}\left(\frac{p}{p-1}1\right)+\frac{7}{2}\left(\frac{p}{p+1}1\right)-\frac{7}{12}\left(\frac{p}{p+2}1\right)+\frac{1}{4}\left(\frac{p}{p-2}1\right)$$

$$=-\frac{7}{6}e^{t}+\frac{7}{2}e^{-t}-\frac{7}{12}e^{-2t}+\frac{1}{4}e^{2t}.$$

The reader will observe that the first three terms belong to the complementary function, and that the last term is a particular solution; he is also invited to check that the initial conditions are satisfied.

Example 2: *Solve* $(D^2-7D+6)x=t^2$,

for the initial conditions $x_0=x_1=0$.

Since $t^2 = \frac{2}{p^2}1$, the operational form of the problem is simply

$$(p^2 - 7p + 6)x = \frac{2}{p^2}1;$$

no other terms appear on the right because the initial values x_0 and x_1 are zero. We now have

$$\frac{x}{p} = \frac{2}{p^3(p-1)(p-6)}1$$
$$= \left(\frac{-2/5}{p-1} + \frac{1/540}{p-6} + \frac{43/108}{p} + \frac{7/18}{p^2} + \frac{1/3}{p^3}\right)1;$$
$$x = \left(-\frac{2}{5}\frac{p}{p-1} + \frac{1}{540}\frac{p}{p-6} + \frac{43}{108} + \frac{7}{18}\frac{1}{p} + \frac{1}{3}\frac{1}{p^2}\right)1$$
$$= -\frac{2}{5}e^t + \frac{1}{540}e^{6t} + \frac{43}{108} + \frac{7}{18}t + \frac{1}{6}t^2.$$

We now give a *rule for solving* (4) *when the operational form of* $f(t)$ *is NOT known* (i.e. when $f(t)$ is not the sum of one or more terms $At^r e^{\alpha t}$):

Write down the operational form

$$A(p)x = pC(p)1 + f(t)$$

of the problem. Then

$$x = p\frac{C(p)}{A(p)}1 + \frac{1}{A(p)}f(t).$$

This is to be evaluated by putting $C(p)/A(p)$ *and* $1/A(p)$ *into partial fractions, and then using formulae* (1) *and* (2).

Note that once again a factor p should be extracted before the first partial fraction expansion (but not before the second one).

Example 3: *Solve* $(D^3 - 3D + 2)x = f(t)$,

for the initial conditions $x_0 = -1$, $x_1 = 5$, $x_2 = 2$.

The operational form, found from the rule on p. 45 or from (5), is

$$(p^3 - 3p + 2)x = (-1(p^3 - 3p) + 5p^2 + 2p)1 + f(t)$$
$$= (-p^3 + 5p^2 + 5p)1 + f(t).$$

Since $p^3-3p+2=(p-1)^2(p+2)$, we obtain

$$x=p\frac{-p^2+5p+5}{(p-1)^2(p+2)}1+\frac{1}{(p-1)^2(p+2)}f(t)$$

$$=p\left(\frac{3}{(p-1)^2}-\frac{1}{p+2}\right)1+\left(\frac{-1/9}{p-1}+\frac{1/3}{(p-1)^2}+\frac{1/9}{p+2}\right)f(t)$$

$$=3te^t-e^{-2t}+\int_0^t\left\{-\frac{1}{9}e^{(t-\tau)}+\frac{1}{3}(t-\tau)e^{(t-\tau)}+\frac{1}{9}e^{-2(t-\tau)}\right\}f(\tau)d\tau.$$

2.3 Some remarks on partial fractions. In the preceding examples we have deliberately suppressed the arithmetical work involved in finding the various partial fraction expansions. We must now warn the reader that this work usually forms a substantial part of the operational method of solution. It is therefore desirable to have a systematic technique for finding partial fractions; we shall now describe such a technique, and shall use it in all future examples.

Suppose then that $G(p)/H(p)$ is a rational function, the degree of the numerator $G(p)$ being less than the degree of the denominator $H(p)$. Let $p-\lambda$ be a typical factor of $H(p)$, with multiplicity m; thus $H(p)=(p-\lambda)^mK(p)$ where $K(p)$ does not contain the factor $p-\lambda$. Then there is an expansion

$$\frac{G(p)}{H(p)}\equiv\frac{A_m}{(p-\lambda)^m}+\frac{A_{m-1}}{(p-\lambda)^{m-1}}+\ldots+\frac{A_1}{p-\lambda}+\ldots, \tag{6}$$

where the last dots represent similar terms corresponding to other factors of $H(p)$. (We assume here that $G(p)/H(p)$ *can* be expressed in the form (6); this fact is proved in textbooks on algebra, e.g. in Durell and Robson, *Advanced Algebra*, Vol. II.) We want to find the coefficients $A_m, \ldots, A_1$ in (6), and in doing this we are permitted to treat p as a numerical variable.

When $p-\lambda$ is a *simple factor* of $H(p)$ (i.e. when $m=1$) the coefficient A in the single corresponding term $A/(p-\lambda)$ can be written down at sight. For then, if we multiply the identity (6) by $H(p)$, we obtain

$$G(p)=AK(p)+(p-\lambda)L(p),$$

where $L(p)$ is a polynomial whose exact form will not concern us. When we put $p=\lambda$ in this identity, we obtain $G(\lambda)=AK(\lambda)$, whence $A=G(\lambda)/K(\lambda)$. Now $K(\lambda)$ can be found either by removing the factor $p-\lambda$ from $H(p)$ and then putting $p=\lambda$, or by differentiating the identity $H(p)\equiv(p-\lambda)K(p)$ and then putting $p=\lambda$, which gives $H'(\lambda)=K(\lambda)$. Hence we may state the following rule:

To find the coefficient A of the term $A/(p-\lambda)$ when $p-\lambda$ is a simple factor of the denominator, strike out the factor $p-\lambda$ from the denominator and then put $p=\lambda$ in the expression which remains. Alternatively, differentiate the denominator with respect to p and then put $p=\lambda$ in the resulting expression $G(p)/H'(p)$.

Usually (but not always) the first version of this rule is more convenient to apply.

When $p-\lambda$ is a *multiple factor* of $H(p)$ (i.e. when $m>1$), we first put $p-\lambda=u$ and express $G(p)/H(p)$ in terms of u; it will then take the form $L(u)/u^mM(u)$ where $L(u)$ and $M(u)$ are again polynomials and $M(u)$ does not contain the factor u (i.e. $M(0)\neq 0$). We now want the coefficients in the expansion

$$\frac{L(u)}{u^mM(u)}\equiv\frac{A_m}{u^m}+\frac{A_{m-1}}{u^{m-1}}+\ldots+\frac{A_1}{u}+\ldots$$

Multiplying up by $u^mM(u)$, we get

$$L(u)\equiv(A_m+A_{m-1}u+\ldots+A_1u^{m-1})M(u)+u^mN(u), \quad (7)$$

where $N(u)$ is a polynomial whose exact form will not concern us. We can now find A_m, A_{m-1}, . . . , A_1 (in that order) by successively equating coefficients for u^0, u^1, . . . , u^{m-1}. The procedure will perhaps be best understood by seeing it at work in a numerical example.

Example 4: Find the terms corresponding to $p-1$ in the partial fractions for $\dfrac{p^3+3p}{(p-1)^3(p+1)(p+2)}$.

Putting $p-1=u$, we have

$$p^3+3p=(u+1)^3+3(u+1)=4+6u+3u^2+u^3$$

and $$(p+1)(p+2)=(u+2)(u+3)=6+5u+u^2.$$

So we want

$$4+6u+3u^2+u^3\equiv(A_3+A_2u+A_1u^2)(6+5u+u^2)+\ldots,$$

the dots representing terms in u^3 and u^4. We can get the correct constant term on the right by choosing A_3 properly; then the correct term in u by choosing A_2, and the correct term in u^2 by choosing A_1. Terms beyond u^2 do not concern us, and in particular the term u^3 on the left has no effect on the values of A_3, A_2 and A_1. The arithmetical work can be performed mentally, but those readers who mistrust their skill in mental arithmetic may prefer to set out the work more systematically like a long division (but done 'in reverse', working in *ascending* powers of u and ignoring terms beyond u^2). Thus

$$\begin{array}{r|l}
 & \frac{2}{3}+\frac{4}{9}u+\frac{1}{54}u^2 \\
6+5u+u^2 & 4+6u+3u^2 \\
 & 4+\frac{10}{3}u+\frac{2}{3}u^2 \\
\hline
 & \frac{8}{3}u+\frac{7}{3}u^2 \\
 & \frac{8}{3}u+\frac{20}{9}u^2 \\
\hline
 & \frac{1}{9}u^2 \\
 & \frac{1}{9}u^2 \\
\hline\hline
\end{array}$$

What this calculations shows is that

$$4+6u+3u^2+u^3 \equiv \left(\frac{2}{3}+\frac{4}{9}u+\frac{1}{54}u^2\right)(6+5u+u^2)+\ldots$$

The fractions which we want are now got by dividing, $\frac{2}{3}+\frac{4}{9}u+\frac{1}{54}u^2$ by u^3 and then replacing u by $p-1$; thus they are

$$\frac{2/3}{(p-1)^3}+\frac{4/9}{(p-1)^2}+\frac{1/54}{p-1}.$$

We may now state the following rule:

To find the terms involving $p-\lambda$ in the partial fractions for $G(p)/H(p)$ when $p-\lambda$ is a multiple factor of the denominator, *put $p-\lambda=u$ and write $G(p)/H(p)$ in the form $L(u)/u^m M(u)$. Then find $A_m+A_{m-1}u+\ldots+A_1u^{m-1}$ from equation (7), divide by u^m, and finally replace u by $p-\lambda$ again. In finding $A_m+\ldots+A_1u^{m-1}$, the method used in Example 4 may be imitated, u^m and higher powers of u being ignored.*

Naturally one must be prepared to meet *complex factors* of the denominator $H(p)$. In practice $G(p)$ and $H(p)$ will usually have real coefficients, so that with any complex factor $p-\lambda$ the conjugate complex factor $p-\bar{\lambda}$ will also occur, and with any fraction $A/(p-\lambda)$ the conjugate fraction $\bar{A}/(p-\bar{\lambda})$. In the evaluation of the solution we will therefore obtain terms like

$$\left(A\frac{p}{p-\lambda}+\bar{A}\frac{p}{p-\bar{\lambda}}\right)1=Ae^{\lambda t}+\bar{A}e^{\bar{\lambda}t}=2\mathbf{R}(Ae^{\lambda t})$$

and these can always be written as real combinations of cosines, sines and exponentials. It must be admitted that multiple complex factors may lead to rather unpleasant calculations, but fortunately they do not occur very often.

2.4 Further examples. [We continue to use the notations x_0, $x_1, \ldots$ for the values of $x, Dx, \ldots$ when $t=0$.]

Example 5: $(D^4+4)x=0$;
initial conditions $x_0=3$, $x_1=x_2=x_3=0$.

The operational form is

$$(p^4+4)x=3p^4 1,$$

so that

$$\frac{x}{p}=\frac{3p^3}{p^4+4}1.$$

Now

$$p^4+4=(p^2-2i)(p^2+2i)=(p-1-i)(p-1+i)(p+1-i)(p+1+i),$$

so that the factors of the denominator are all simple. If $p-\lambda$ is any one of these factors, the coefficient of the corresponding term in the partial fractions for $\dfrac{3p^3}{p^4+4}$ may be found by differentiating the denominator and then putting $p=\lambda$ (see the rule on p. 49 which gives

$$\left[\frac{3p^3}{4p^3}\right]_{p=\lambda}=\frac{3}{4}$$

for all four choices of λ. Hence

$$x=\frac{3}{4}\left(\frac{p}{p-1-i}+\frac{p}{p-1+i}+\frac{p}{p+1-i}+\frac{p}{p+1+i}\right)1$$

$$=\frac{3}{2}\mathbf{R}(e^{(1+i)t}+e^{(-1+i)t})$$

$$=\frac{3}{2}e^t\cos t+\frac{3}{2}e^{-t}\cos t=3\cosh t\cos t.$$

Example 6: $(D-1)^3x=\sin t$;

initial conditions $x_0=1,\ x_1=2,\ x_3=3$.

The operational form of $\sin t$ can be found thus:

$$\sin t=\mathbf{I}(e^{it})=\mathbf{I}\left(\frac{p}{p-i}1\right)=\mathbf{I}\left(\frac{p(p+i)}{p^2+1}1\right)=\frac{p}{p^2+1}1.$$

Also $(p-1)^3=p^3-3p^2+3p-1$. So the operational form of the problem is

$$(p-1)^3x=\left(1(p^3-3p^2+3p)+2(p^2-3p)+3(p)+\frac{p}{p^2+1}\right)1$$

$$=\left(p^3-p^2+\frac{p}{p^2+1}\right)1=\frac{p^5-p^4+p^3-p^2+p}{p^2+1}1,$$

whence

$$\frac{x}{p}=\frac{p^4-p^3+p^2-p+1}{(p-1)^3(p^2+1)}1.$$

The complex factors $p\pm i$ in the denominator give rise to two conjugate fractions, and the coefficient of $1/(p-i)$ is

$$\frac{i^4-i^3+i^2-i+1}{(i-1)^3 2i}=\frac{1}{(2+2i)2i}=-\frac{1}{4(1-i)}=-\frac{1+i}{8}.$$

The corresponding contribution to x is $-\frac{1+i}{8}e^{it}$ plus the conjugate term, that is

$$2\mathbf{R}\left\{-\frac{1+i}{8}(\cos t+i\sin t)\right\}=-\tfrac{1}{4}(\cos t-\sin t).$$

For the terms corresponding to the multiple factor $p-1$, put $p-1=u$. This gives

$$\frac{p^4-p^3+p^2-p+1}{(p-1)^3(p^2+1)}=\frac{1+2u+4u^2+\ldots}{u^3(2+2u+u^2)},$$

dots denoting terms in u^3 and u^4 which we shall not need. Following the method used in Example 3, we obtain

$$\begin{aligned}1+2u+4u^2+\ldots &= \tfrac{1}{2}(2+2u+u^2)\\ &\quad+\tfrac{1}{2}u(2+2u+u^2)\\ &\quad+\frac{5}{4}u^2(2+2u+u^2)+\ldots\end{aligned}$$

Thus the required terms are obtained by dividing $\frac{1}{2}+\frac{1}{2}u+\frac{5}{4}u^2$ by u^3 and replacing u by $p-1$; the corresponding contribution to x is

$$\left(\frac{1}{2}\frac{p}{(p-1)^3}+\frac{1}{2}\frac{p}{(p-1)^2}+\frac{5}{4}\frac{p}{p-1}\right)1.$$

Evaluating this, and adding on the contribution found earlier, we get the solution

$$x=-\frac{1}{4}(\cos t-\sin t)+\left(\frac{1}{4}t^2+\frac{1}{2}t+\frac{5}{4}\right)e^t.$$

Example 7: $(D^2+\omega^2)x=f(t)$ ($\omega\neq 0$ *being a real constant*)*; arbitrary initial conditions* $x=x_0$, $Dx=x_1$.

The operational form is

$$(p^2+\omega^2)x=(x_0p^2+x_1p)1+f(t).$$

Hence

$$x=p\frac{x_0p+x_1}{p^2+\omega^2}1+\frac{1}{p^2+\omega^2}f(t).$$

The required partial fractions, easily found, are

$$\frac{x_0p+x_1}{p^2+\omega^2}=\frac{\tfrac{1}{2}\left(x_0-i\dfrac{x_1}{\omega}\right)}{p-i\omega}+\frac{\tfrac{1}{2}\left(x_0+i\dfrac{x_1}{\omega}\right)}{p+i\omega},\quad \frac{1}{p^2+\omega^2}=\frac{1}{2i\omega}\left(\frac{1}{p-i\omega}-\frac{1}{p+i\omega}\right),$$

so that we get

$$\begin{aligned}x&=2\mathrm{R}\left\{\tfrac{1}{2}\left(x_0-i\frac{x_1}{\omega}\right)e^{i\omega t}\right\}+\frac{1}{2i\omega}\int_0^t\{e^{i\omega(t-\tau)}-e^{-i\omega(t-\tau)}\}f(\tau)d\tau\\ &=x_0\cos\omega t+\frac{x_1}{\omega}\sin\omega t+\frac{1}{\omega}\int_0^t\sin\omega(t-\tau)f(\tau)d\tau.\end{aligned}$$

Example 8: $(D^2+2D+5)x=e^{-2t}$;

initial conditions $x_0=-3$, $x_1=2$.

The operational form is

$$(p^2+2p+5)x=\left(-3(p^2+2p)+2(p)+\frac{p}{p+2}\right)1;$$

also $p^2+2p+5=(p+1-2i)(p+1+2i)$. Hence we have

$$\frac{x}{p}=-\frac{3p^2+10p+7}{(p+2)(p+1-2i)(p+1+2i)}1.$$

In the partial fraction expansion

$$-\frac{3p^2+10p+7}{(p+2)(p+1-2i)(p+1+2i)}\equiv\frac{A}{p+2}+\frac{B}{p+1-2i}+\frac{\bar{B}}{p+1+2i},$$

we have

$$A=\left[-\frac{3p^2+10p+7}{p^2+2p+5}\right]_{p=-2}=-\frac{12-20+7}{4-4+5}=+\frac{1}{5},$$

$$B=\left[-\frac{3p^2+10p+7}{(p+2)(p+1+2i)}\right]_{p=-1+2i}=-\frac{-12+8i}{(1+2i)(4i)}=\frac{-8+i}{5}.$$

Hence

$$x=\left(A\frac{p}{p+2}+B\frac{p}{p+1-2i}+\bar{B}\frac{p}{p+1+2i}\right)1$$

$$=\frac{1}{5}e^{-2t}+2\mathrm{R}\left(\frac{-8+i}{5}e^{(-1+2i)t}\right)$$

$$=\frac{1}{5}e^{-2t}+\frac{2}{5}e^{-t}\mathrm{R}((-8+i)(\cos 2t+i\sin 2t))$$

$$=\frac{1}{5}e^{-2t}-\frac{2}{5}e^{-t}(8\cos 2t+\sin 2t).$$

Example 9: $(D^4+2D^2+1)x=0$;

initial conditions $x_0=1$, $x_1=0$, $x_2=-1$, $x_3=0$.

The operational form is

$$(p^4+2p^2+1)x=(1(p^4+2p^2)-1(p^2))1=(p^4+p^2)1.$$

We are lucky because the factor (p^2+1) can be cancelled, and

$$x=p\frac{p}{p^2+1}1=p\frac{2}{2}\left(\frac{1}{p-i}+\frac{1}{p+i}\right)1=\frac{1}{2}\left(\frac{p}{p-i}+\frac{p}{p+i}\right)1.$$

So the solution

$$x=\tfrac{1}{2}(e^{it}+e^{-it})=\cos t.$$

Example 10: $(D^4+2D^2+1)x=0$;

initial conditions $x_0=x_1=x_2=0$, $x_3=1$.

This time we have $(p^4+2p^2+1)x=p1$, and $\dfrac{x}{p}=\dfrac{1}{(p^2+1)^2}1$. The denominator is $(p-i)^2\,(p+i)^2$; the terms in $p-i$ are found by putting $p-i=u$, which gives

$$\frac{1}{(p^2+1)^2}=\frac{1}{u^2(u+2i)^2}=\frac{1}{u^2(-4+4iu+\ldots)}.$$

Now $1=(-\tfrac{1}{4}-\tfrac{1}{4}iu)(-4+4iu+\ldots)+\ldots$, dots denoting terms in u^2. Dividing $-\tfrac{1}{4}-\tfrac{1}{4}iu$ by u^2 and replacing u by $p-i$, we find that

$x=-\tfrac{1}{4}\dfrac{p}{(p-i)^2}1-\tfrac{1}{4}i\dfrac{p}{p-i}1$ plus the conjugate terms. Hence

$$\begin{aligned} x&=2\mathrm{R}\{-\tfrac{1}{4}te^{it}-\tfrac{1}{4}ie^{it}\}\\ &=-\tfrac{1}{2}t\cos t+\tfrac{1}{2}\sin t. \end{aligned}$$

We have deliberately set out the solutions to examples in such a way that only the formulae (1) and (2) from p. 42 need be used. More elaborate formulae could sometimes be used to shorten the work slightly, the most useful being

$$\frac{p(p-\alpha)}{(p-\alpha)^2+\omega^2}1=e^{\alpha t}\cos\omega t,\quad \frac{\omega p}{(p-\alpha)^2+\omega^2}1=e^{\alpha t}\sin\omega t. \tag{8}$$

We do not advise the reader to make strenuous efforts to memorize (8), since it can be derived very simply by taking real and imaginary parts in

$$\frac{p}{p-\alpha-i\omega}1=e^{\alpha t}e^{i\omega t}.$$

2.5 Simultaneous equations. These can also be treated by the operational method. First consider the simplest problem: to solve two first order equations

$$\left.\begin{aligned}(a_1D+b_1)x+(c_1D+d_1)y=f(t),\\ (a_2D+b_2)x+(c_2D+d_2)y=g(t),\end{aligned}\right\} \tag{9}$$

subject to the initial conditions

$$x=x_0,\ y=y_0,\ \text{when } t=0. \tag{10}$$

The equivalent pair of equations involving Q is found by integrating equations (9), which gives

$$a_1(x-x_0)+b_1Qx+c_1(y-y_0)+d_1Qy=Qf(t),$$
$$a_2(x-x_0)+b_2Qx+c_2(y-y_0)+d_2Qy=Qg(t).$$

Now put $Q=p^{-1}$, and multiply by p, obtaining

$$\left.\begin{aligned}(a_1p+b_1)x+(c_1p+d_1)y=(x_0a_1+y_0c_1)p1+f(t),\\(a_2p+b_2)x+(c_2p+d_2)y=(x_0a_2+y_0c_2)p1+g(t).\end{aligned}\right\}\qquad(11)$$

This is the *operational form* of the problem; if the operational forms of $f(t)$ and $g(t)$ are known, we of course insert these. The equations (11) are now solved by ordinary algebra, p being treated like a number, and the result is evaluated in the usual way.

Example 11:

$$\left.\begin{aligned}(D-1)x-2y=t,\\-2x+(D-1)y=t;\\\textit{initial conditions } x_0=2,\ y_0=4.\end{aligned}\right\}$$

The operational form is

$$(p-1)x-2y=2p1+\frac{1}{p}1,$$

$$-2x+(p-1)y=4p1+\frac{1}{p}1.$$

Solving for x,

$$[(p-1)^2-4]x=\left((p-1)\left(2p+\frac{1}{p}\right)+2\left(4p+\frac{1}{p}\right)\right)1,$$

$$\frac{x}{p}=\frac{2p^3+6p^2+p+1}{p^2(p+1)(p-3)}1.$$

After finding partial fractions in the usual way, we obtain

$$x=p\left(-\frac{1/3}{p^2}-\frac{1/9}{p}-\frac{1}{p+1}+\frac{28/9}{p-3}\right)1$$
$$=-\frac{1}{3}t-\frac{1}{9}-e^{-t}+\frac{28}{9}e^{3t}.$$

We could find y similarly, but it is slightly simpler to note that

$$2y=Dx-x-t=-\frac{1}{3}+e^{-t}+\frac{28}{3}e^{3t}+\frac{1}{3}t+\frac{1}{9}+e^{-t}-\frac{28}{9}e^{3t}-t,$$

whence $y=-\frac{1}{3}t-\frac{1}{9}+e^{-t}+\frac{28}{9}e^{3t}.$

In exceptional cases the above treatment may break down. Suppose that in equations (9) the terms containing derivatives, a_1Dx+c_1Dy and a_2Dx+c_2Dy, are proportional. It is then possible to obtain a single equation, involving x and y but not their derivatives. For example, if the equations are

$$Dx+2x+Dy-3y=t,$$
$$2Dx+x+2Dy+2y=0,$$

then we can derive the single equation

$$3x-8y=2t.$$

In such a case, the initial values x_0 and y_0 can certainly not be independently assigned (in the above example we must have $3x_0-8y_0=0$), and even when x_0 and y_0 are suitably connected there may be no solution to the problem. We shall not treat such cases; they should not occur in physical applications, provided that the physical problem has been correctly translated into mathematical terms.

The treatment of n simultaneous first order equations follows lines which should now be obvious. Denote the unknowns by $x_1, \ldots, x_n$; let the equations read

$$\sum_{s=1}^{n}(a_{rs}D+b_{rs})x_s=f_r(t) \quad (r=1, 2, \ldots, n), \tag{12}$$

and let the initial conditions be

$$x_r=\alpha_r \text{ when } t=0 \quad (r=1, 2, \ldots, n). \tag{13}$$

Then the operational form of the problem reads

$$\sum_{s=1}^{n}(a_{rs}p+b_{rs})x_s=\sum_{s=1}^{n} a_{rs}\alpha_s p1+f_r(t) \quad (r=1, \ldots, n), \tag{14}$$

and these equations are to be solved by ordinary algebra. The solution can then be evaluated in the usual way. It is assumed that the terms in (12) which involve derivatives are independent, i.e. that it is impossible to combine these equations so as to obtain a single equation free from derivatives. This assumption is equivalent[1] to the following one:

[1] See P. M. Cohn, *Linear Equations*, Ch. V, § 12 (in this series).

The determinant formed from the coefficients a_{rs} $(r, s=1, \ldots, n)$ *is not zero,*

and it ensures that the initial values $\alpha_1, \ldots, \alpha_n$ may be independently assigned.

More complicated systems of equations can also be dealt with. We content ourselves with one illustration: suppose we have two second order equations in unknowns x and y. Let x_0, x_1, y_0, y_1 denote the values of x, Dx, y, Dy, when $t=0$, and let the first of the two equations read

$$(aD^2+bD+c)x+(a'D^2+b'D+c')y=f(t).$$

Then the corresponding operational equation can be written down by an obvious extension of the rule (p. 45) for a single second order equation:

on the left, replace D *by* p*; on the right, add to* $f(t)$ *the terms*

$$(x_0(ap^2+bp)+x_1(ap)+y_0(a'p^2+b'p)+y_1(a'p))1.$$

The second equation is treated similarly, and the resulting pair of operational equations is then solved by ordinary algebra.

Example 12:

$$\left.\begin{aligned}(D^2-4D)x-(D-1)y&=1,\\ (D+6)x+(D^2-D)y&=e^{4t};\\ \textit{initial conditions } x_0=x_1=y_0&=1,\ y_1=0.\end{aligned}\right\}$$

The operational form of the problem is

$$\begin{aligned}(p^2-4p)x-(p-1)y&=(1(p^2-4p)+1(p)+1(-p)+1)1\\ &=(p^2-4p+1)1,\\ (p+6)x+(p^2-p)y&=\left(1(p)+1(p^2-p)+0(p)+\frac{p}{p-4}\right)1\\ &=\left(p^2+\frac{p}{p-4}\right)1.\end{aligned}$$

Solving for x,

$$\begin{aligned}[(p^2-4p)(p^2-p)+(p-1)(p+6)]x&=(p^2-p)(p^2-4p+1)1\\ &\quad+(p-1)\left(p^2+\frac{p}{p-4}\right)1\\ &=\left(p^4-4p^3+4p^2-p+\frac{p(p-1)}{p-4}\right)1.\end{aligned}$$

The coefficient of x on the left is

$$p^4-5p^3+5p^2+5p-6=(p+1)(p-1)(p-2)(p-3);$$

on the right we have

$$\frac{p}{p-4}(p^4-8p^3+20p^2-16p+3)1=\frac{p(p-1)(p-3)(p^2-4p+1)}{p-4}1.$$

Hence we get

$$\frac{x}{p}=\frac{p^2-4p+1}{(p+1)(p-2)(p-4)}1=\left(\frac{2/5}{p+1}+\frac{1/2}{p-1}+\frac{1/10}{p-4}\right)1,$$

$$x=\frac{2}{5}e^{-t}+\frac{1}{2}e^{2t}+\frac{1}{10}e^{4t}.$$

To find y, we have

$$\begin{aligned}(p-1)y&=p(p-4)x-(p^2-4p+1)1\\&=(p^2-4p+1)\left(\frac{p^2}{(p+1)(p-2)}-1\right)1\\&=(p^2-4p+1)\frac{p+2}{(p+1)(p-2)}1,\end{aligned}$$

$$\frac{y}{p}=\frac{(p^2-4p+1)(p+2)}{p(p+1)(p-1)(p-2)}1=\left(\frac{1}{p}-\frac{1}{p+1}+\frac{3}{p-1}-\frac{2}{p-2}\right)1,$$

$$y=1-e^{-t}+3e^{t}-2e^{2t}.$$

Once again, difficulties arise when the terms involving second derivatives are not independent. For two equations this means simply that the second derivative terms are proportional; in the example, these terms were D^2x and D^2y, and no trouble occurred because these terms are independent.

2.6 Justification of the method. We have not yet supplied a detailed proof of the facts that the operational method, applied either to an nth order equation (§ 2.2) or to a system of first order equations (§ 2.5), gives the correct solution to initial value problems. This gap will now be filled; the proofs will be given in terms of Q (more convenient than p in theoretical work).

Consider first the problem of solving the equations (12) on p. 57, subject to the initial conditions (13). If $x_1, \ldots, x_n$ are a solution to this problem then by integrating (12) and using the fact that $QDx_s(t) = x_s(t) - x_s(0) = x_s(t) - \alpha_s$ we obtain

$$\sum_{s=1}^{n} a_{rs}(x_s - \alpha_s) + \sum_{s=1}^{n} b_{rs}Qx_s = Qf_r(t),$$

$$\left.\begin{aligned}\sum_{s=1}^{n}(a_{rs}+b_{rs}Q)x_s &= \sum_{s=1}^{n} a_{rs}\alpha_s + Qf_r(t)\\ &= g_r(t), \text{ say } (r=1, \ldots, n).\end{aligned}\right\} \quad (15)$$

Conversely, if $x_1, \ldots, x_n$ satisfy (15) then (differentiating) they satisfy the differential equations (12). Also, putting $t=0$ in (15), the terms involving Q vanish and we get

$$\sum_{s=1}^{n} a_{rs}(x_s(0) - \alpha_s) = 0 \quad (r=1, \ldots, n).$$

But we are assuming that the determinant of the a_{rs} is non-zero, and therefore these equations imply that $x_r(0) - \alpha_r = 0$, i.e. that the initial conditions (13) are satisfied. Thus we have shown that the initial value problem (given by (12) and (13)) is equivalent to the set of equations (15). Note that when we put $Q = p^{-1}$ in (15) and multiply through by p, we obtain precisely equations (14), i.e. the equations which are actually used in practical work (where p is more convenient than Q).

We now show that the equations (15) have exactly one solution. Denote $a_{rs}+b_{rs}Q$ by $c_{rs}(Q)$, let $\Delta(Q)$ be the determinant whose $(r, s)^{\text{th}}$ element is $c_{rs}(Q)$ and let $C_{rs}(Q)$ be the cofactor of $c_{rs}(Q)$. Then $\Delta(Q)$ is a polynomial of degree at most n in Q, say

$$\Delta(Q) = \Delta_0 + \Delta_1 Q + \ldots + \Delta_n Q^n,$$

in which the constant term Δ_0 is simply the determinant of the a_{rs} and is therefore different from zero; also $C_{rs}(Q)$ is a polynomial in Q. (We are using determinants whose elements are operators; they can be defined and treated like numerical determinants because their elements all commute, so that multiplications can be performed in any order.) Now let $x_1, \ldots, x_n$ be any solution of (15). Multiply the rth equation in (15) by $C_{rj}(Q)$, and add; then the coefficient of x_s on the left is

$$\sum_{r=1}^{n} C_{rj}(Q)c_{rs}(Q) = \begin{cases}\Delta(Q) \text{ when } s=j.\\ 0 \text{ when } s \neq j.\end{cases}$$

So only x_j survives, and we must have

$$\Delta(Q)x_j = \sum_{r=1}^{n} C_{rj}(Q)g_r(t) \qquad (j=1, \ldots, n).$$

The operator $\Delta(Q)$ on the left has the form $\Delta_0(1+k_1Q+\ldots+k_nQ^n)$ with $\Delta_0 \neq 0$; it therefore has an inverse (cf. § 1.5). It follows that we must have

$$x_j = \sum_{r=1}^{n} \frac{C_{rj}(Q)}{\Delta(Q)} g_r(t) \qquad (j=1, \ldots, n). \tag{16}$$

A similar argument, which we omit, will show that (16) does give a solution to (15); this completes the proof that (15) has exactly one solution, given by (16). Of course all we have done in the above proof is to go through one of the methods of solving linear equations with numerical coefficients and to check that the steps are still valid when the coefficients are operators of the form $a_{rs}+b_{rs}Q$ the resulting formula (16) for the solution is 'Cramer's rule'.[1] This justifies the second step in the operational method, which consists in solving (15) 'by ordinary algebra'. The final step is to evaluate the solution (16), which involves putting rational functions of Q into partial fractions; this has already been justified in § 1.5. We have now established the following fact:

the differential equations (12) *have exactly one solution satisfying given initial conditions of the form* (13), *provided that the determinant formed from the coefficients a_{rs} is different from zero, and this solution can be calculated by the operational method as set out in* § 2.5 (*in its practical form, using p rather than Q*).

We turn now to the initial value problem for a single n^{th} order equation, stated as (4) on p. 44. To justify the operational method of solving this problem, we shall replace it by an equivalent problem for a set of n first order equations. Let us take x, $Dx, \ldots, D^{n-1}x$ as new unknowns and denote them by $u_0, u_1, \ldots, u_{n-1}$. Then we have $Du_0=u_1$, $Du_1=u_2, \ldots, Du_{n-2}=u_{n-1}$, and the differential equation for x (in (4)) becomes $Du_{n-1}+a_1u_{n-1}+\ldots+a_nu_0=f(t)$; also $u_0, \ldots, u_{n-1}$ have the initial values $x_0, \ldots, x_{n-1}$. Conversely if $u_0, \ldots, u_{n-1}$ satisfy the preceding equations and initial conditions, then $x=u_0$ is a solution to problem (4). So an

[1] See P. M. Cohn, *Linear Equations*, Ch. V, §§ 12–13 (in this series).

initial value problem equivalent to (4) is given by

$$Du_0 - u_1 = 0,\ Du_1 - u_2 = 0,\ \ldots,\ Du_{n-2} - u_{n-1} = 0,$$
$$Du_{n-1} + a_1 u_{n-1} + a_2 u_{n-2} + \ldots + a_n u_0 = f(t); \quad (17)$$
$$u_0 = x_0,\ u_1 = x_1,\ \ldots,\ u_{n-1} = x_{n-1} \text{ when } t = 0. \quad (18)$$

But this, when inessential differences in notation are disregarded, is precisely the type of problem stated in (12) and (13); the crucial determinant, formed from the coefficients of the derivative terms, is non-zero because its elements are 1 on the main diagonal and 0 elsewhere. So we may write down the operational form of the problem as in (15), namely

$$\left.\begin{aligned} &u_0 - Qu_1 = x_0,\ u_1 - Qu_2 = x_1,\ \ldots,\ u_{n-2} - Qu_{n-1} = x_{n-2}, \\ &u_{n-1} + a_1 Qu_{n-1} + a_2 Qu_{n-2} + \ldots + a_n Qu_0 = x_{n-1} + Qf(t). \end{aligned}\right\} \quad (19)$$

We may then solve these equations by ordinary algebra; only the solution for $u_0 (= x)$ is wanted, and this can be obtained by multiplying the equations (19) by

$$\begin{aligned} &1 + a_1 Q + \ldots + a_{n-1} Q^{n-1}, \\ &Q + a_1 Q^2 + \ldots + a_{n-2} Q^{n-1}, \\ &\cdot \quad \cdot \quad \cdot \quad \cdot \quad \cdot \\ &Q^{n-2} + a_1 Q^{n-1}, \\ &Q^{n-1}, \end{aligned}$$

and adding. Writing x for u_0, this gives

$$(1 + a_1 Q + a_2 Q^2 + \ldots + a_n Q^n) x = Q^n f(t)$$
$$+ \left\{\begin{aligned} &x_0(1 + a_1 Q + \ldots + a_{n-1} Q^{n-1}) \\ &+ x_1(Q + a_1 Q^2 + \ldots + a_{n-2} Q^{n-1}) + \ldots + x_{n-1} Q^{n-1} \end{aligned}\right\} 1, \quad (20)$$

and x can now be found because the operator $1 + a_1 Q + \ldots + a_n Q^n$ has an inverse. This justifies the operational method in its theoretical version, using Q. It only remains to observe that when we replace Q by p^{-1} in (20) and multiply through by p^n, we obtain precisely the operational form (5) of problem (4) which is used in the practical version (using p) of the operational method. We have therefore established that

the initial value problem (4) *has exactly one solution, and this can be calculated by the operational method as set out in* § 2.2.

2.7 The general solution of an nth order equation. The structure of the general solution of

$$(D^n + a_1 D^{n-1} + \ldots + a_n) x = f(t), \quad (21)$$

becomes very clear when the operational method is applied. Whatever initial conditions are chosen, we obtain a solution of the form

$$x = p\frac{C(p)}{A(p)}1 + \frac{1}{A(p)}f(t), \tag{22}$$

where $A(p) = p^n + a_1 p^{n-1} + \ldots + a_n$ and the polynomial $C(p)$ depends on the initial conditions but does not depend on $f(t)$. Thus (22) displays the solution as the sum of two terms; the first term, $p\frac{C(p)}{A(p)}1$, satisfies the reduced equation

$$(D^n + \ldots + a_n)x = 0 \tag{23}$$

with the same initial conditions as before, whilst the second term, $\frac{1}{A(p)}f(t)$, is the particular solution of (21) for which the initial values of $x, Dx, \ldots, D^{n-1}x$ are all zero. Moreover we see that *every* solution of the reduced equation has the form $p\frac{C(p)}{A(p)}1$; when we evaluate this expression, after factorizing $A(p)$ as

$$(p-\lambda_1)^{m_1}(p-\lambda_2)^{m_2} \ldots (p-\lambda_k)^{m_k},$$

we obtain a linear combination of

$$e^{\lambda_1 t}, te^{\lambda_1 t}, \ldots, t^{m_1-1}e^{\lambda_1 t}; \ldots; e^{\lambda_k t}, \ldots, t^{m_k-1}e^{\lambda_k t} \tag{24}$$

Conversely, as we already showed in Ch. I (p. 21), any linear combination of the functions (24) is a solution of the reduced equation. We have therefore confirmed what was stated without proof on p. 22:

> *the general solution of the reduced equation* (23) *is an arbitrary linear combination of the functions* (24).

Now suppose that we merely want to find the general solution of (21). Then all we need, after writing down the complementary function, is a particular solution. The operational method at once provides us with one particular solution, namely $\frac{1}{A(p)}f(t)$. We can do even better when the operational form $F(p)1$ of $f(t)$ is known, so that we get the particular solution

$\frac{F(p)}{A(p)}1$, because when we evaluate this (by partial fraction expansion) we may omit all terms which have already occurred in the complementary function. For example, a particular solution of $(D^4-1)x=te^t$ is

$$\frac{1}{p^4-1}\frac{p}{(p-1)^2}1=p\frac{1}{(p^2+1)(p+1)(p-1)^3}1,$$

but when we make the expansion

$$\frac{1}{(p^2+1)(p+1)(p-1)^3}=\frac{A}{p-i}+\frac{\bar{A}}{p+i}+\frac{B}{p+1}+\frac{C_1}{p-1}+\frac{C_2}{(p-1)^2}+\frac{C_3}{(p-1)^3}$$

we need only the last two terms because the other terms will merely produce multiples of e^{it}, e^{-it}, e^{-t} and e^t, which are already in the complementary function. The technique for partial fractions described in § 2.3 is very convenient here: it enables us to isolate a particular group of terms without having to find the other (unwanted) terms. The numerical work involved is perhaps no shorter than if one were to substitute a trial solution, but has the advantage of following a familiar pattern when one has become used to the operational method.

PROBLEMS FOR CHAPTER II

[The values of x, Dx, D^2x, . . . at $t=0$ will be denoted by x_0, x_1, x_2, . . . ; D stands for d/dt.]

1. Check the partial fractions occurring in Examples 1–3 of § 2.2.
2. $(D^2+6D+9)x=t^2e^{-3t}$; $x_0=x_1=0$.
3. $(D^2-D-2)x=e^t \sin 2t$; $x_0=x_1=0$.
4. $(D^4-5D^3+5D^2+5D-6)x=0$; $x_0=x_3=2$, $x_1=x_2=0$.
5. $(D^3+3D^2-D-3)x=e^{-2t}$; $x_0=x_1=x_2=0$.
6. $(D^2+2D+2)x=f(t)$; $x_0=x_1=0$.
7. $(D^2-2D+1)x=e^t$; $x_0=1$, $x_1=-1$.
8. $(D^3+1)x=e^{-t}$; $x_0=1$, $x_1=-1$, $x_2=-2$.
9. $(D^2-5D+6)x=\cos 3t$; $x_0=0$, $x_1=5$.

10. $(D^3-3D^2+3D-1)x=54e^{4t}$; $x_0=1$, $x_1=2$, $x_2=3$.
11. $(D^4+2D^3+D^2-2D-2)\,x=t$; $x_0=1$, $x_1=-2$, $x_2=0$, $x_3=5$.
12. $(D^4-5D^3+6D^2+4D-8)x=0$; $x_0=0$, $x_1=x_2=1$, $x_3=3$.
13. $(D^4+4)x=f(t)$; $x_0=x_1=x_2=x_3=0$.
14. $(D^4+4D^2+4)x=1+t^2$; $x_0=x_1=x_2=x_3=0$.
15. $(D^3-D^2+4D-4)x=15e^t$; $x_0=x_1=5$, $x_2=0$.
16. $\left.\begin{array}{l}(3D+2)x+Dy=0,\\ Dx+(4D+3)y=17e^t,\end{array}\right\}x_0=y_0=0.$
17. $\left.\begin{array}{l}(D^2-8)x+Dy=0,\\ -6Dx+(D^2+2)y=0,\end{array}\right\}x_0=y_0=1,\ x_1=-1,\ y_1=6.$
18. $\left.\begin{array}{l}(D^2+5)x-y=4t^2,\\ -3x+(D^2+3)y=0,\end{array}\right\}x_0=x_1=y_0=y_1=0.$
19. $D^2x+2x-z=D^2y+2y-z=D^2z-8x-y+2z=0$;
$x_0=x_1=y_0=y_1=z_1=0$, $z_0=1$. Find z.
20. $\left.\begin{array}{l}(2D+3)x+Dy=f(t),\\ Dx+(D+2)y=g(t),\end{array}\right\}x_0=y_0=0.$ Find x.

SOLUTIONS TO PROBLEMS

[Arbitrary constants will be called A, B, C, sometimes with suffixes attached.]

Chapter I (p. 25):

1. $Ce^x-(x^2+2x+2)$.
2. xe^x.
3. $\frac{2}{5}\cos x+\frac{1}{5}\sin x+Ce^{-2x}$.
4. $e^{-x}(C+\log(1+e^{2x}))$.
5. $e^{-2x}(2+A\cos x+B\sin x)$.
6. $\frac{1}{5}e^{-x}(\cos x-2\sin x)+A\cos x+B\sin x$.
7. $\frac{1}{5}\cos x+Ae^{2x}+Be^{3x}$.
8. $-\frac{1}{4}x^2+\frac{3}{4}x+\frac{7}{8}+Ae^{2x}+Be^{-2x}$.
9. $(A+Bx+\frac{1}{6}x^3)e^{-x}$.
10. $\frac{1}{10}x(2\sin x-\cos x)+\frac{1}{50}(\cos x-7\sin x)$.
[Replace R.H.S. by xe^{ix}, substitute trial solution $(Ax+B)e^{ix}$, then take the imaginary part.]
11. $\frac{5}{2}e^{-x}-\frac{1}{2}e^{3x}$.
12. $\frac{1}{2}x(\cos x+3\sin x)$.
13. $\frac{2}{3}e^{2x}+Ae^{-x}+e^{\frac{1}{2}x}\left(B\cos\frac{\sqrt{3}}{2}x+C\sin\frac{\sqrt{3}}{2}x\right)$.
14. $2\cos 3x-\frac{3}{2}\sin 3x+Ae^{-x}+e^{2x}(B\cos x+C\sin x)$.

15. $x^3+12x+(C_1x+C_2)e^x+(C_3x+C_4)e^{-x}$.
16. $y=3e^x-e^{-x},\ z=3e^x-3e^{-x}$.
17. $y=e^{-x}(A\cos x+B\sin x)$,
$z=e^{-x}(A\sin x-B\cos x)$.

Chapter II (p. 64):

2. $\frac{1}{12}t^4e^{-3t}$. **3.** $\frac{2}{15}e^{2t}-\frac{1}{12}e^{-t}-\frac{1}{20}e^t(\cos 2t+3\sin 2t)$.
4. $\frac{7}{2}e^t+\frac{5}{12}e^{-t}-\frac{8}{3}e^{2t}+\frac{3}{4}e^{3t}$. **5.** $\frac{1}{24}e^t-\frac{1}{4}e^{-t}+\frac{1}{3}e^{-2t}+\frac{1}{8}e^{-3t}$.

6. $\int_0^t e^{-(t-\tau)}\sin(t-\tau)f(\tau)d\tau$. **7.** $e^t(1-2t+\frac{1}{2}t^2)$.

8. $\frac{1}{3}(1+t)e^{-t}+e^{\frac{1}{2}t}\left(\frac{2}{3}\cos\left(\frac{t\sqrt{3}}{2}\right)-\frac{8}{3\sqrt{3}}\sin\left(\frac{t\sqrt{3}}{2}\right)\right)$.

9. $-\frac{67}{13}e^{2t}+\frac{31}{6}e^{3t}-\frac{1}{78}(\cos 3t+5\sin 3t)$.
10. $2e^{4t}-(1+5t+9t^2)e^t$.
11. $\frac{1}{2}(1-t)+e^{-t}(\frac{5}{2}\cos t-\sin t-2)$. **12.** $\frac{1}{3}(e^{2t}-e^{-t})$.

13. $\frac{1}{4}\int_0^t\{\cosh(t-\tau)\sin(t-\tau)-\cos(t-\tau)\sinh(t-\tau)\}f(\tau)d\tau$.

14. $\frac{1}{4}(t^2-1+\cos(t\sqrt{2}))$.
15. $(3t+\frac{14}{5})e^t+\frac{1}{5}(11\cos 2t-2\sin 2t)$.
16. $x=\frac{1}{10}(-5e^t+17e^{-t}-12e^{-\frac{6}{11}t}),\ y=\frac{1}{10}(25e^t-17e^{-t}-8e^{-\frac{6}{11}t})$.
17. $x=\frac{1}{4}e^{2t}+\frac{1}{2}e^{-2t}+\frac{1}{4}(\cos 2t-\sin 2t)$,
$y=\frac{1}{2}e^{2t}-e^{-2t}+\frac{3}{2}(\cos 2t+\sin 2t)$.
18. $x=t^2-\frac{2}{3}+\frac{1}{2}\cos(t\sqrt{2})+\frac{1}{6}\cos(t\sqrt{6})$,
$y=t^2-\frac{4}{3}+\frac{3}{2}\cos(t\sqrt{2})-\frac{1}{6}\cos(t\sqrt{6})$.
19. $z=\frac{1}{2}\cosh t+\frac{1}{2}\cos(t\sqrt{5})$.

20. $\frac{1}{5}\int_0^t e^{-(t-\tau)}(f(\tau)+g(\tau))d\tau-\frac{2}{5}\int_0^t e^{-6(t-\tau)}(3g(\tau)-2f(\tau))d\tau$.

Index

LIBRARY OF MATHEMATICS

Edited by W Ledermann

Complex Numbers	W Ledermann
Differential Calculus	P J Hilton
Differential Geometry	K L Wardle
Electrical and Mechanical Oscillations	D S Jones
Elementary Differential Equations and Operators	G E H Reuter
Fourier and Laplace Transforms	P D Robinson
Fourier Series	I N Sneddon
Functions of a Complex Variable I and II	D O Tall
Integral Calculus	W Ledermann
Introduction to Abstract Algebra	C R J Clapham
Linear Equations	P M Cohn
Linear Programming	Kathleen Trustrum
Multiple Integrals	W Ledermann
Numerical Approximation	B R Morton
Partial Derivatives	P J Hilton
Principles of Dynamics	M B Glauert
Probability Theory	A M Arthurs
Sequences and Series	J A Green
Sets and Groups	J A Green
Solid Geometry	P M Cohn
Solutions of Laplace's Equation	D R Bland
Vibrating Strings	D R Bland
Vibrating Systems	R F Chisnell

LIBRARY OF MATHEMATICS

Edited by Walter Ledermann

The aim of this series is to provide short introductory text-books for the topics which are normally covered in the first two years of mathematics courses at Universities and Colleges of Technology. Each volume is made as nearly self-contained as possible, with exercises and answers, and contains an amount of material that can be covered in about twenty lectures. Thus each student will be able to build up a collection of text-books which is adapted to the syllabus he has to follow.

The exposition is kept at an elementary level with due regard to modern standards of rigour. When it is not feasible to give a complete treatment, because this would go beyond the scope of the book, the assumptions are fully explained and the reader is referred to appropriate works in the literature.

'The authors obviously understand the difficulties of undergraduates. Their treatment is more rigorous than what students will have been used to at school, and yet it is remarkably clear.

'All the books contain worked examples in the text and exercises at the ends of the chapters. They will be invaluable to undergraduates. Pupils in their last year at school, too, will find them useful and stimulating. They will learn the university approach to work they have already done, and will gain a foretaste of what awaits them in the future.' – *The Times Educational Supplement*

'It will prove a valuable corpus. A great improvement on many works published in the past with a similar objective.' – *The Times Literary Supplement*

'These are all useful little books, and topics suitable for similar treatment are doubtless under consideration by the edtior of the series.' – T. A. A. BROADBENT, *Nature*

A complete list of books in the series appears on the inside back cover.

ROUTLEDGE & KEGAN PAUL